NATIONAL ACADEMIES
Sciences
Engineering
Medicine

NATIONAL ACADEMIES PRESS
Washington, DC

Cryptography and the Intelligence Community

The Future of Encryption

Committee on the Future of Encryption

Intelligence Community Studies Board

Division on Engineering and Physical Sciences

Consensus Study Report

THE NATIONAL ACADEMIES PRESS 500 Fifth Street, NW Washington, DC 20001

This study was supported by Contract No. 2020-20011300401-002 with the Office of the Director of National Intelligence. Any opinions, findings, conclusions, or recommendations expressed in this publication do not necessarily reflect the views of any agency or organization that provided support for the project.

International Standard Book Number-13: 978-0-309-49135-8
International Standard Book Number-10: 0-309-49135-5
Digital Object Identifier: https://doi.org/10.17226/26168

Copies of this publication are available from

Intelligence Community Studies Board
National Academies of Sciences, Engineering, and Medicine
500 Fifth Street, NW, Room 928
Washington, DC 20001

This publication is available from the National Academies Press, 500 Fifth Street, NW, Keck 360, Washington, DC 20001; (800) 624-6242 or (202) 334-3313; http://www.nap.edu.

Copyright 2022 by the National Academy of Sciences. National Academies of Sciences, Engineering, and Medicine and National Academies Press and the graphical logos for each are all trademarks of the National Academy of Sciences. All rights reserved.

Printed in the United States of America.

Suggested citation: National Academies of Sciences, Engineering, and Medicine. 2022. *Cryptography and the Intelligence Community: The Future of Encryption*. Washington, DC: The National Academies Press. https://doi.org/10.17226/26168.

The **National Academy of Sciences** was established in 1863 by an Act of Congress, signed by President Lincoln, as a private, nongovernmental institution to advise the nation on issues related to science and technology. Members are elected by their peers for outstanding contributions to research. Dr. Marcia McNutt is president.

The **National Academy of Engineering** was established in 1964 under the charter of the National Academy of Sciences to bring the practices of engineering to advising the nation. Members are elected by their peers for extraordinary contributions to engineering. Dr. John L. Anderson is president.

The **National Academy of Medicine** (formerly the Institute of Medicine) was established in 1970 under the charter of the National Academy of Sciences to advise the nation on medical and health issues. Members are elected by their peers for distinguished contributions to medicine and health. Dr. Victor J. Dzau is president.

The three Academies work together as the **National Academies of Sciences, Engineering, and Medicine** to provide independent, objective analysis and advice to the nation and conduct other activities to solve complex problems and inform public policy decisions. The National Academies also encourage education and research, recognize outstanding contributions to knowledge, and increase public understanding in matters of science, engineering, and medicine.

Learn more about the National Academies of Sciences, Engineering, and Medicine at **www.nationalacademies.org**.

Consensus Study Reports published by the National Academies of Sciences, Engineering, and Medicine document the evidence-based consensus on the study's statement of task by an authoring committee of experts. Reports typically include findings, conclusions, and recommendations based on information gathered by the committee and the committee's deliberations. Each report has been subjected to a rigorous and independent peer-review process and it represents the position of the National Academies on the statement of task.

Proceedings published by the National Academies of Sciences, Engineering, and Medicine chronicle the presentations and discussions at a workshop, symposium, or other event convened by the National Academies. The statements and opinions contained in proceedings are those of the participants and are not endorsed by other participants, the planning committee, or the National Academies.

Rapid Expert Consultations published by the National Academies of Sciences, Engineering, and Medicine are authored by subject-matter experts on narrowly focused topics that can be supported by a body of evidence. The discussions contained in rapid expert consultations are considered those of the authors and do not contain policy recommendations. Rapid expert consultations are reviewed by the institution before release.

For information about other products and activities of the National Academies, please visit www.nationalacademies.org/about/whatwedo.

COMMITTEE ON THE FUTURE OF ENCRYPTION

STEVEN B. LIPNER, NAE,[1] SAFECode/Carnegie Mellon University, *Chair*
MARK LOWENTHAL, Intelligence & Security Academy, LLC, *Vice Chair*
HANS ROBERT DAVIES, Toffler Associates
CHIP ELLIOTT, BBN Technologies
GLENN S. GERSTELL, Center for Strategic & International Studies
NADIA HENINGER, University of California, San Diego
SENY KAMARA, Brown University
PAUL CARL KOCHER, NAE, American Cryptographer
BRIAN LaMACCHIA, Microsoft Research
BUTLER W. LAMPSON, NAS[2]/NAE, Microsoft Research
RAFAIL OSTROVSKY, University of California, Los Angeles
ELIZABETH RINDSKOPF PARKER, State Bar of California (retired)
PETER SWIRE, Georgia Institute of Technology
PETER J. WEINBERGER, Google, Inc.

Staff

CARYN A. LESLIE, Acting Director, Intelligence Community Studies Board
JON EISENBERG, Director, Computer Science and Telecommunications Board
LYNETTE MILLETT, Senior Program Officer, Computer Science and Telecommunications Board
MARGUERITE SCHNEIDER, Administrative Coordinator, Intelligence Community Studies Board

---

[1] Member, National Academy Engineering.
[2] Member, National Academy of Sciences.

INTELLIGENCE COMMUNITY STUDIES BOARD

MARK LOWENTHAL, Intelligence & Security Academy, LLC, *Co-Chair*
MICHAEL A. MARLETTA, NAS[1]/NAM,[2] University of California, Berkeley, *Co-Chair*
JOEL BRENNER, Massachusetts Institute of Technology
ROBERT CARDILLO, The Cardillo Group, LLC
FREDERICK R. CHANG, NAE,[3] Southern Methodist University
DEAN CHENG, The Heritage Foundation
ROBERT C. DYNES, NAS, University of California (president emeritus)
ROBERT A. FEIN, Harvard Medical School
HUBAN A. GOWADIA, Lawrence Livermore National Laboratory
MARGARET A. HAMBURG, NAM, Nuclear Threat Initiative
MIRIAM E. JOHN, Independent Consultant
ANITA K. JONES, NAE, University of Virginia (professor emerita)
STEVEN E. KOONIN, NAS, Center for Urban Science and Progress
CARMEN L. MIDDLETON, The Walt Disney Company
ARTHUR L. MONEY, NAE, Department of Defense
WILLIAM C. OSTENDORFF, United States Naval Academy
DAVID A. RELMAN, NAM, Stanford University
ELIZABETH RINDSKOPF PARKER, State Bar of California (retired)
SAMUEL S. VISNER, The MITRE Corporation
DAVID A. WHELAN, NAE, Cubic

Staff

DIONNA ALI, Associate Program Officer
BRYAN BUNNELL, Research Associate
JOSEPH CZIKA, Senior Program Officer
MICHAEL ANTHONY FAINBERG, Senior Program Officer
CARYN A. LESLIE, Acting Director
NIA JOHNSON, Program Officer
MARGUERITE SCHNEIDER, Administrative Coordinator

---

[1] Member, National Academy of Sciences.
[2] Member, National Academy of Medicine.
[3] Member, National Academy of Engineering.

Acknowledgment of Reviewers

This Consensus Study Report was reviewed in draft form by individuals chosen for their diverse perspectives and technical expertise. The purpose of this independent review is to provide candid and critical comments that will assist the National Academies of Sciences, Engineering, and Medicine in making each published report as sound as possible and to ensure that it meets the institutional standards for quality, objectivity, evidence, and responsiveness to the study charge. The review comments and draft manuscript remain confidential to protect the integrity of the deliberative process.

We thank the following individuals for their review of this report:

Cynthia Beall, NAS,[1] Case Western Reserve University,
Thomas A. Berson, NAE,[2] Salesforce,
Susan Landau, The Fletcher School and Tufts School of Engineering,
Marvin J. Langston, Langston Associates, LLC,
John Manferdelli, VMWare,
Julie J.H.C. Ryan, Wyndrose Technical Group, and
Fred Schneider, NAE, Cornell University.

Although the reviewers listed above provided many constructive comments and suggestions, they were not asked to endorse the conclusions or recommendations of this report nor did they see the final draft before its release. The review of this report was overseen by Robert F. Sproull, NAE, University of Massachusetts Amherst, and Deborah Westphal, Toffler Associates. They were responsible for making certain that an independent examination of this report was carried out in accordance with the standards of the National Academies and that all review comments were carefully considered. Responsibility for the final content rests entirely with the authoring committee and the National Academies.

---

[1] Member, National Academy of Sciences.
[2] Member, National Academy of Engineering.

Preface

The U.S. Intelligence Community, like intelligence organizations worldwide, uses encryption to protect sensitive information from disclosure and modification, and also seeks to decrypt encrypted information that it collects as part of its mission. In 2020, the Office of the Director of National Intelligence requested that the National Academies of Sciences, Engineering, and Medicine conduct a study to explore the future of encryption over the next 10 to 20 years. The study was to explore technical and non-technical drivers that would affect the viability of the community's use of encryption to protect information and the challenges of defeating adversaries' encryption and to produce a set of scenarios that would illustrate possible futures in which the Intelligence Community would have to operate.

The National Academies established the Committee on the Future of Encryption to conduct the study. The full statement of task for the committee is shown in Appendix A. The biographies of the committee members that authored this report are shown in Appendix B.

Committee members included academics, industrial researchers, and engineering practitioners in cryptography and computer science as well as attorneys and policy and intelligence professionals. They brought great expertise in the technology of encryption, its applications and integration into information systems and networks, and the policies and operations of government agencies that both use and seek to defeat encryption. Because of the constraints posed by the COVID-19 pandemic, the committee was unable to meet in person and held virtual meetings biweekly from September 2020 to September 2021.

The committee operated under the auspices of the National Academies' Intelligence Community Studies Board and is grateful for the able assistance of Caryn A. Leslie, Marguerite Schneider, and Lynette Millett of the National Academies' staff.

Contents

SUMMARY 1

1 INTRODUCTION 12

2 INTRODUCTION TO ENCRYPTION 16

3 METHODOLOGY 35

4 DRIVERS 40

5 SCENARIOS 81

6 IMPLICATIONS FOR U.S. INTELLIGENCE 101

7 FINDINGS 105

APPENDIXES

A	Statement of Task	113
B	Meeting Agendas	114
C	Potential Scenarios	116
D	*Global Trends 2040*	119
E	Acronyms and Abbreviations	122
F	Committee Member Biographical Information	125

Summary

CRYPTOGRAPHY AND THE INTELLIGENCE COMMUNITY

Encryption is a process for making information unreadable by an adversary who does not possess a specific key that is required to make the encrypted information readable. The inverse process, making information that has been encrypted readable, is referred to as decryption. Encryption and decryption are facets of a broad scientific field referred to as cryptography.[1] (See Box S.1 for definitions.[2]) For most of recorded history, encryption was an arcane process used primarily by governments, the military, and a few commercial organizations that sought to protect their communications from disclosure.[3] Today, cryptography has become widespread and is used by private as well as governmental actors. It also enables authentication (verifying the identities of people, software, and the origins of transactions), and underlies the safe use of the Internet and computer systems by individuals and organizations worldwide. Emerging cryptographic technologies offer capabilities such as the ability to process encrypted information without first decrypting it. Cryptography is a complex and specialized subject: Chapter 2 of the body of this report introduces aspects of cryptography for readers who are not familiar with it, and the National Institute of Standards and Technology's (NIST's) glossary is a useful reference.[4]

The U.S. Intelligence Community, like intelligence organizations worldwide, uses encryption to protect sensitive information from unauthorized disclosure or modification, and it also has to decrypt encrypted information that it collects as part of its mission. The protective use of encryption is referred to as "defensive" and the task of defeating encryption as "offensive." The defensive role of the U.S. Intelligence Community extends to setting standards and/or creating systems for the encryption of classified U.S. national security information and advising on the creation of standards for the encryption of unclassified government and private sector information. The offensive role involves the collection of intelligence about the activities of governments and non-state actors that pose a potential threat to the interests of the United States.

The Office of the Director of National Intelligence (ODNI) requested that the National Academies of Sciences, Engineering, and Medicine establish a committee to identify potential scenarios that would describe the

---

[1] D. Boneh and V. Shoup, 2020, "A Graduate Course in Applied Cryptography," Version 0.5, January, http://toc.cryptobook.us.

[2] J. Katz and Y. Lindell, 2021, *Introduction to Modern Cryptography*, Boca Raton, FL: Chapman & Hall/CRC Press, Taylor & Francis Group.

[3] D. Khan, 1967, *The Codebreakers: The Story of Secret Writing*, New York: Macmillan.

[4] National Institute of Standards and Technology (NIST), 2019, *Glossary of Key Information Security Terms*, NISTIR 7298 Revision 3, https://nvlpubs.nist.gov/nistpubs/ir/2019/NIST.IR.7298r3.pdf.

BOX S.1
Encryption Concepts and Terminology

Private-Key Encryption

The simplest form of encryption is called private-key encryption or symmetric encryption. To keep information (called plaintext) secret, the sender encrypts it with a key to obtain ciphertext; using the same key, the recipient can decrypt the ciphertext and recover the original plaintext, but without it the ciphertext reveals no meaningful information about the plaintext. Figure S.1.1 illustrates the operation of private-key encryption.

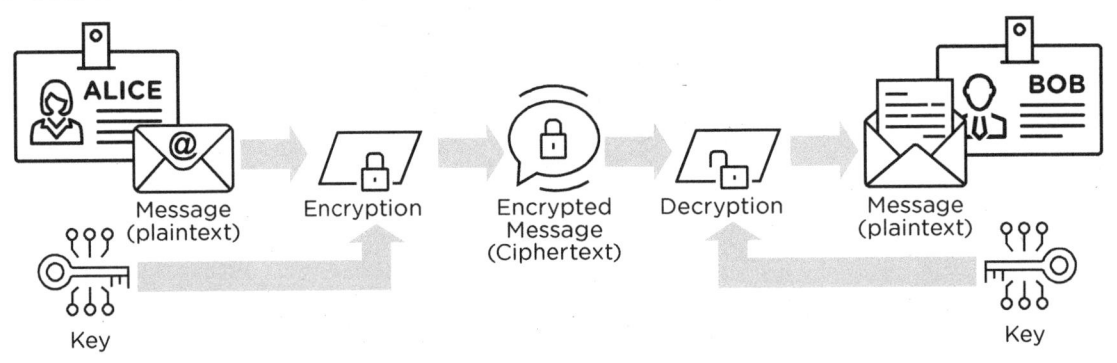

FIGURE S.1.1 Private-key encryption.

Public-Key Encryption

While symmetric, or private-key, encryption uses the same key for both encryption and decryption operations, public-key or asymmetric encryption uses different but mathematically related pairs of keys for these two functions. Critically, in a public-key system knowledge of the encryption key does not imply knowledge of the corresponding decryption key. This means that if a message recipient publishes the encryption key and keeps its matching decryption key secret, anyone can send messages that only the recipient can decrypt and read. Figure S.1.2 illustrates the operation of public-key encryption.

SUMMARY

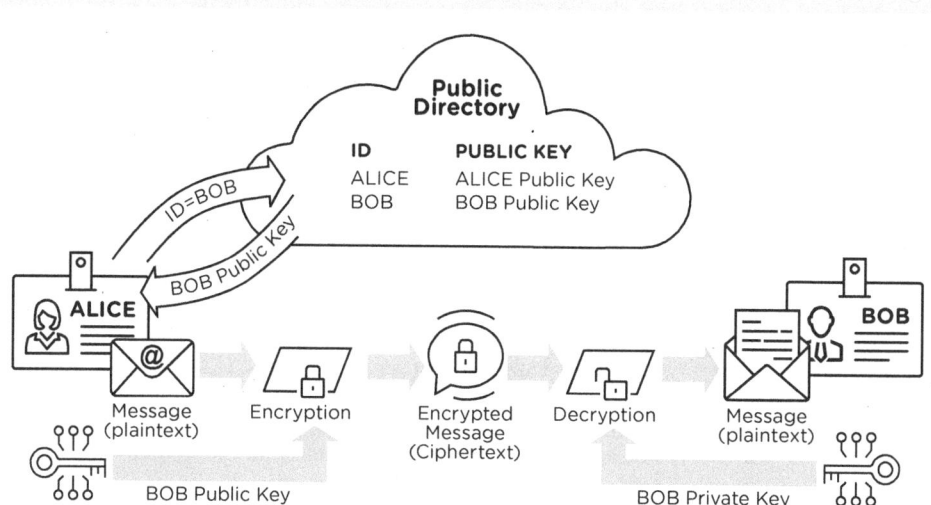

FIGURE S.1.2 Public-key encryption.

Digital Signatures

Digital signature schemes use mathematical techniques similar to those used in public-key encryption schemes, but with the specific purpose of authenticating data. A digital signature scheme allows anyone in possession of the public verification key to verify a digital signature of a message but only someone in possession of the matching private signing key can sign the message. Figure S.1.3 illustrates the operation of digital signature systems.

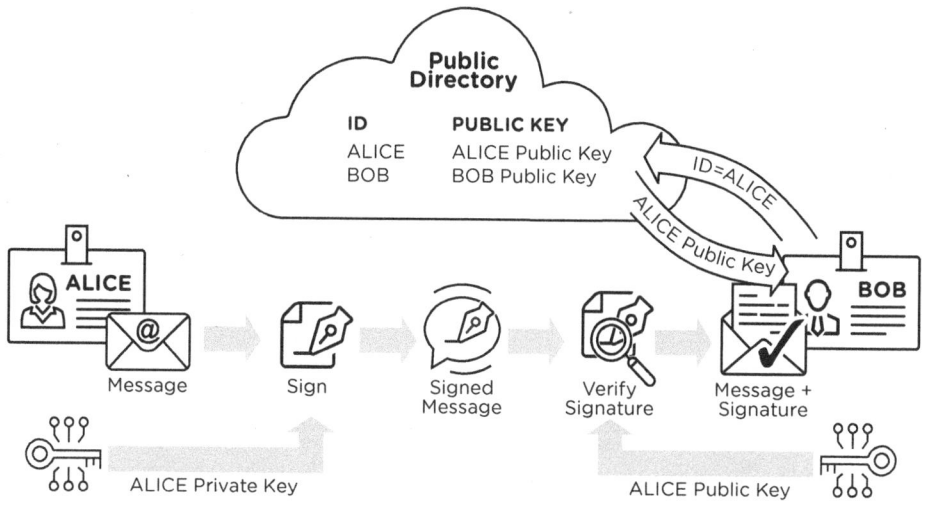

FIGURE S.1.3 Digital signatures.

balance between encryption and decryption over the next 10 to 20 years and to assess the national security and intelligence implications of each scenario. The committee's objective is to identify the range of possible developments and their implications, and to provide the Intelligence Community with guidance on ways of identifying which future scenarios are most likely to emerge so that the Intelligence Community and the U.S. government as a whole can respond to and take advantage of these changes. Given the wide range of possible futures, the report does not attempt to predict what specific developments will occur or become dominant over the next 10 to 20 years or specifically when those developments might occur.

The committee's work dealt only with cryptographic systems and technologies described by public sources and did not involve access to classified information. Given that governments and national security organizations worldwide are major users of commercial and public systems (including their security mechanisms), dealing with these systems and technologies is a significant consideration and challenge for the Intelligence Community.

TRENDS AND MOTIVATIONS

The committee's work is well motivated by the importance of cryptography to the Intelligence Community and by the scope and variety of encryption-related changes in government policies and commercial technology. However, one technical issue stands out as a particular motivator—the potential development of quantum computers—computational systems that would rely on phenomena of quantum physics to perform computation in a fundamentally different way from the computers that have been built since the 1940s. Research results[5] have shown that a sufficiently large-scale working quantum computer could be programmed to defeat current asymmetric[6] (or public-key) encryption systems that are fundamental to the security of the Internet.[7,8]

Researchers, governments, and industry are well aware of the potential impact of quantum computers, and work is under way to identify, standardize, and implement "post-quantum" asymmetric encryption systems that are believed not to be subject to attack by a quantum computer.[9] However, the transition from current encryption systems to post-quantum encryption systems will require the replacement of a vast amount of software and some hardware that is fundamental to the operation of Internet-connected computer systems. The potential impact of quantum computers and the implications of the transition to post-quantum encryption systems were major subjects of the committee's work.

IDENTIFYING SCENARIOS TO DESCRIBE THE FUTURE OF ENCRYPTION

The statement of task (see Appendix A) requires the committee to identify scenarios for the future of encryption and potential areas of technology surprise. Alternative scenarios result from combinations of technical and other "drivers" that influence the direction of technology as well as the decisions and actions of individuals and governments.

To identify scenarios of interest to the Intelligence Community, the committee used a mix of approaches commonly found in the work of futurists, strategic foresight firms, and academic researchers. The committee first identified technical and non-technical drivers whose possible future states are important to the future of encryption. The committee then used combinations of the extreme endpoint values of those future states to define potential

---

[5] P.W. Shor, 1997, Polynomial-time algorithms for prime factorization and discrete logarithms on a quantum computer, *SIAM Journal on Computing* 26(5):1484–1509, https://doi.org/10.1137/s0097539795293172.

[6] Unlike encryption systems whose origins go back millennia, asymmetric encryption systems apply a public key that can be shared widely to encrypt information, and a separate private key, related to the public key by a hard-to-solve mathematical problem, to decrypt information. Most websites and services on the Internet rely on asymmetric encryption to achieve user authentication and data protection.

[7] S. Goldwasser and S. Micali, 1984, Probabilistic encryption, *Journal of Computer and System Sciences* 28(2):270–299, https://doi.org/10.1016/0022-0000(84)90070-9.

[8] National Academies of Sciences, Engineering, and Medicine, 2019, *Quantum Computing: Progress and Prospects*, Washington, DC: The National Academies Press, https://doi.org/10.17226/25196.

[9] NIST, "Post-Quantum Cryptography," Computer Security Resource Center, Information Technology Laboratory, https://csrc.nist.gov/projects/post-quantum-cryptography, accessed October 12, 2021.

scenarios. This report focuses on the potential scenarios that the committee determined to be most informative and to cover the broadest range of plausible futures. Once those potential scenarios had been identified, the committee assessed their implications for the Intelligence Community and considered potential actions in response to each selected scenario.

DRIVERS AND SCENARIOS

The Intelligence Community confronts a future of encryption that will result from developments that span technology, the dynamics of society and policy, and the marketplace. Specific technical breakthroughs may have significant impact on the effectiveness of encryption or on intelligence organizations' abilities to defeat encryption or both. The ability and willingness of organizations to trust allied governments and their own employees will also have a significant impact. Last, organizations' ability and willingness to create and operate robust, reliable products will also affect the effectiveness of the encryption and decryption capabilities upon which the Intelligence Community relies.

The committee identified three major drivers that it believes will greatly influence the future of encryption over the next 10 to 20 years:

- *Scientific Advances:* The emergence of new theoretical breakthroughs or significant technologies that affect cryptography. The creation of a large-scale quantum computer might be one such advance. Others include new mathematical attacks on asymmetric encryption, advances that enable efficient computation on encrypted data, and technologies that use quantum properties for encryption. The endpoints of the state of future scientific advances could be either predictable or disruptive.
- *Society and Governance:* The policies, politics, and points of view that influence the way cryptography is used and the way that individuals and organizations apply and manage it. The actions of governments could bear heavily on the future use of cryptography: government commitments to protect their citizens' privacy might be implemented by requiring that all personal information about citizens be encrypted and/or processed only in encrypted form. Governments might also try to enforce restrictions on citizens' use of encryption. Or governments might withdraw from intelligence-sharing agreements with other nations that have historically supported offensive attacks on adversaries' encryption. Individuals' trust and confidence in their own governments might also erode, leading citizens to seek encryption solutions that would protect their information from their own governments or making it difficult for governments to find trustworthy people to create, operate, and manage their encryption systems. The two endpoints of the future of society and governance could be either *global* or *fragmented*.
- *Systems:* The soundness of the products and technology that implement, embed, or support encryption. Products and network protocols that incorporate encryption could be correct, reliable, and "bug free," or they could be laden with design and implementation errors that require constant patching and that would undermine security, even if the mathematical theories underlying the basic encryption mechanisms are sound. For example, developers might integrate otherwise sound cryptographic building blocks in a way that makes a final system insecure, or they might create software that fails to protect encryption keys. At the endpoints, future cryptographic systems could be either *mature* or *chaotic*.

SCENARIOS FOR THE FUTURE OF ENCRYPTION

As noted above, the committee considered the two endpoints for each of the three drivers. Although no actual future will manifest precisely according to the outcome represented by any of the combinations of endpoints, focusing on these extremes enabled the committee to explore plausible outcomes and their potential impacts. Focusing on attributes of the endpoints will also enable the Intelligence Community to identify, calibrate, and evaluate the observable trends that are shaping the future of encryption.

Taken together, the combinations of the endpoints of the three drivers define eight possible scenarios for the future of encryption.[10] The committee selected three of those eight for in-depth exploration based on the objectives of covering especially plausible futures and of exploring scenarios that resulted in the most challenges and opportunities for the Intelligence Community. These chosen scenarios, the endpoint descriptions of their associated drivers, and brief descriptions of each scenario are included in Table S.1.

TABLE S.1 Scenario Descriptions

Scenario Title	Driver Endpoints	Highlight
A Brave and Expensive New World	Disruptive / Fragmented / Mature	A breakthrough in quantum computing is balanced with more secure systems and software and an orderly transition to post-quantum encryption.
Scenario Description		
This scenario posits that a breakthrough in quantum computing is offset by an orderly transition to post-quantum encryption and other emerging cryptographic techniques, because of earlier investments in systems and cybersecurity. Overall, the balance now favors defense. However, the global political picture remains fragmented. The bottom line for the Intelligence Community is that offensive cryptography efforts have become more difficult, and the alliance structure that is a major plus for U.S. intelligence is less reliable and more fluid. In this scenario, a major issue for U.S. intelligence will be its ability to discern, far enough in advance, the development of a reliable, large-scale quantum computer. This is of crucial importance to this scenario, which posits the development of more secure systems because governments and the private sector take seriously the need for improved cybersecurity and invest in the development and deployment of much more robust systems. Even if such a quantum computing breakthrough did not occur, there are obvious benefits in enhanced systems and cybersecurity. If these steps are taken early enough and on a wide basis, then the transition to a post-quantum world would be orderly. It will be important for U.S. intelligence to be able to discern in advance the development of a reliable large-scale quantum computer both to ensure that progress toward post-quantum encryption is sufficient for defensive purposes and to assess potential offensive opportunities. International relations in this scenario fragment into a small number of blocs composed of a few major powers (the United States, the European Union, China) and a large number of dependent powers that, for the most part, rely on a major power for economic, technological, and defense support. Technology and encryption are similarly fragmented, with blocs sharing common approaches to encryption and inter-bloc communications relying on weak "least common denominator" encryption. Within blocs, government surveillance is the norm, although some governments and some private companies use new capabilities for processing encrypted data to protect sensitive processing.		

Scenario Title	Driver Endpoints	Highlight
The Known World, Only More So	Predictable / Fragmented / Chaotic	With no major breakthroughs and a continued lack of focus on systems and security, breaches remain common; meanwhile, the slow pace of technology change has allowed emerging competitors the chance to "catch up."
Scenario Description		
This scenario posits that there are no major breakthroughs regarding a quantum computer, as well as a continued lack of focus on systems and security. Therefore, system breaches remain common. Also, the slow pace of technology change has allowed emerging competitors the chance to "catch up." The overall balance continues to favor offense. The key issue under this scenario for U.S. intelligence is the broadening of the threat. Already the world is one in which many states and a growing number of non-state actors pose threats to U.S. and allied interests. These threats will likely increase in number and severity in this scenario, potentially compounded by weakening of traditional alliances and partnerships. At the same time, a world of this sort remains a "target rich" environment for U.S. intelligence collection efforts. The constantly growing Internet of Things (IoT) also adds to this threat and to the opportunity. There is also the question of what tools, if any, the Intelligence Community should develop to respond to/retaliate against such attacks. Policies and decisions to do so belong to the policy community, but the Intelligence Community should be prepared to offer a range of responsive options. Success in this scenario requires a very serious assessment of U.S.—meaning government at all levels and the private sector—vulnerabilities, in order to take steps to ameliorate or eliminate them. Given that this is the largely known world at present, it may be difficult to motivate people and organizations to make changes that appear to be expensive and time consuming for perhaps modest gains in security. Private sector firms may also be wary, if not suspicious, about government efforts to foster changes. The measures in the Biden administration's executive order are largely voluntary. Legislation may be necessary (like seat belt and speed limit laws) to foster real change. The Intelligence Community could be asked to give advice as to what measures are needed, although this sort of participation by U.S. intelligence is likely to foster greater suspicion about backdoors or other means of great government intrusion. Engaging the Department of Homeland Security (DHS) and the National Institute of Standards and Technology (NIST) in the discussion might make recommendations more broadly acceptable and allay some of these concerns about intrusion.		

---

[10] See Appendix C for a summary of the eight possible scenarios.

TABLE S.1 Continued

Scenario Title	Driver Endpoints	Highlight
Colony Collapse	Disruptive / Fragmented / Chaotic	A breakthrough in factoring and a lack of focus on cryptographic systems and security puts information at risk. Despite advances in computing on encrypted data, public trust remains low.

Scenario Description

This scenario posits a breakthrough in the form of a new classical factoring algorithm. Such a breakthrough would render the public key encryption algorithms currently relied on more easily attacked with much less effort than today, including by conventional computers. In this case, there would be much less need for a quantum computer. A roomful of powerful servers might be sufficient. Compared to a quantum computing breakthrough, a factoring breakthrough would probably have less advance notice, be easier to keep secret, and be attainable by more countries.

In addition, a lack of focus on systems and security puts information at risk in this scenario. Despite advances in computing on encrypted data, trust remains low. This suggests, overall, a much more chaotic world combined with, as in the other scenarios, a more fragmented world politically.

In this scenario, the Intelligence Community has to be on the lookout for a breakthrough that will be less easily discerned and more easily hidden than a large-scale quantum computer. Such a breakthrough could come via the efforts of a government or from academic cryptographic researchers. In either case, there might not be advance notice that such a breakthrough was about to occur or had occurred. Experts will debate the likelihood of such a factoring breakthrough, but it seems prudent to ask what would happen should it occur and what steps would need to be taken at that point. As in the other scenarios, much would depend on the ability of the U.S. government to mobilize the private sector to take the necessary steps as well.

As with the other scenarios, international fragmentation—of technical standards as well as governmental relations—complicates the challenges for the Intelligence Community. Many governments react to internal tensions and fragmentation by weakening or imposing limits on the use of encryption, further facilitating attacks on encryption.

In some respects, the results in this scenario are potentially more widespread and more dangerous than in the other two scenarios, giving added urgency to examinations of the likelihood and consequences of this breakthrough happening.

COMMON TRENDS AND KEY FINDINGS

As directed by the statement of task, the committee identified trends that are common across some or all scenarios and potential responses to those trends. This report documents these trends in findings that identify risks, opportunities, and actions. Attention to the findings should enable the Intelligence Community to prepare for the future and to recognize emerging trends and developments and respond appropriately. Chapter 7 enumerates the full set of findings, risks, opportunities, and actions.

This section presents several key findings that result from overarching considerations and trends. These considerations and trends span most or all scenarios and are especially important to the future that the Intelligence Community and society will likely face.

Chaotic Systems Are Likely to Undermine the Security of Encryption

The wide dependence of governments, companies, and individuals on commercial information-technology products amplifies the importance of those products' implementation and applications of encryption. This dependence is reflected in the Systems Driver.

Today, the state of the Systems Driver is chaotic. Vulnerabilities, bugs, and other errors are frequent. Those errors can often enable attackers to bypass or undermine encryption.[11,12] Such vulnerabilities have been reported in

---

[11] D. Bleichenbacher, 1998, "Chosen Ciphertext Attacks Against Protocols Based on the RSA Encryption Standard PKCS #1," pp. 1–12 in *Advances in Cryptology—CRYPTO '98*, https://doi.org/10.1007/bfb0055716.

[12] R. Cramer and V. Shoup, 1998, "A Practical Public Key Cryptosystem Provably Secure Against Adaptive Chosen Ciphertext Attack," pp. 13–25 in *Advances in Cryptology—CRYPTO '98*, https://doi.org/10.1007/bfb0055717.

hardware and in both commercial and open source cryptographic software and can compromise the confidentiality and/or integrity of information.[13,14]

The transition to post-quantum encryption will be delayed by chaotic systems, and when the transition is complete, chaotic systems may have made the resulting products less trustworthy. To the extent that this situation does not change—and in the committee's view there is little reason for optimism—it has an extremely significant effect in enabling offense (attacking cryptography) to prevail over defense. In other words, breaking systems will be easier than protecting them.

It is possible that some countries, vendors, or open-source projects could achieve mature systems while others do not, but the market has not historically been kind to products or companies that prioritized security over performance, usability, and time to market. In response to this situation, a recent U.S. government executive order seeks to push vendors to create more mature systems, but it is far too soon to predict its impact.

Key Findings

FINDING 4.8: In every scenario, bugs in software and other issues outside the underlying cryptographic algorithms and protocols are the weakest links in security.[15]

FINDING 4.9: Communications and storage depend on a software stack: hypervisor (a program that allows a computer to run several operating systems simultaneously), operating system, libraries, and application. While quantum computers or mathematical advances are important research topics, bugs or operational mistakes in this stack are the biggest source of system insecurity. Exploiting these errors is, and likely will remain, the biggest opportunity for offense, and minimizing them the highest priority for defense and risk management.

FINDING 4.13: The complexity of the transition to post-quantum cryptography will likely introduce a range of new security vulnerabilities.

FINDING 4.15: 5G may introduce a number of new systems issues in practice, owing to both complex new suites of software and operator inexperience in distributed cloud environments.

FINDING 4.16: Many Internet of Things (IoT) components are poorly secured and easy to subvert, with an extremely wide range of consequences that are difficult to predict but potentially very high impact for the Intelligence Community and broader society. Because IoT will likely bring significant improvements to many aspects of life, however, more money and energy may be devoted to securing such devices going forward.

Fragmented Society and Governance Are Likely to Degrade the Security of Systems and Organizations That Rely on Encryption

The breakdown of international cooperation that is likely to follow from fragmented society and governance will have significant effects on the trustworthiness and effectiveness of encryption. The recent Global Trends report produced by the National Intelligence Council, released as the committee finalized its work, depicts scenarios that are almost all fragmented.[16] The committee found that distrust among nations and groups, and the breakdown of national alliances, will lead to wider use of cryptography, less sharing of information (and, in particular, less

---

[13] C. Cimpanu, 2020, "Microsoft Fixes Windows Crypto Bug Reported by the NSA," ZDNet, January 14, https://www.zdnet.com/article/microsoft-fixes-windows-crypto-bug-reported-by-the-nsa.

[14] Wikipedia, 2021, "Heartbleed," Wikimedia Foundation, July 10, https://en.wikipedia.org/wiki/Heartbleed.

[15] T. Armerding, 2016, "The OPM Breach Report: A Long Time Coming," CSO Online, October 13, https://www.csoonline.com/article/3130682/the-opm-breach-report-a-long-time-coming.html.

[16] The *Global Trends 2040* report (National Intelligence Council, 2021, *Global Trends 2040: A More Contested World*, Office of the Director of National Intelligence, March, https://www.dni.gov/index.php/gt2040-home) is summarized in Appendix D.

sharing of intelligence), and the proliferation of a variety of exclusive national post-quantum encryption standards. Many of these trends are likely to make the offensive task more challenging.

Fragmented society and governance will also affect citizens' trust in and loyalty to their governments. Degradation in trust and loyalty favors offense, making potential targets of those individuals who are trusted with access to cryptographic systems, mechanisms, and keys. The potential introduction of law enforcement access mandates intended to defeat encryption under specific circumstances is a double-edged sword. It may provide law enforcement agencies with expanded access (an offensive advantage) but has been criticized by many in the technical community because of the likelihood that it will introduce a defensive weakness in the form of a target for adversaries who would seek to use the available law enforcement access mechanisms in unauthorized and unintended ways.

Key Finding

FINDING 4.6: Governmental regulation, for better or worse, of communications technology may lead to fragmentation on national lines. National security concerns have the effect, whether specifically intended or not, of creating competing national technologies—by limiting the exports of sensitive technology or by curtailing imports of equipment that may permit surreptitious surveillance by a foreign manufacturer or its government. Potent forces are present, for both beneficial and malicious reasons that could predispose the global arrangement toward individual nationalistic or regional solutions to issues bearing on encryption. In many countries, there is growing support for "digital sovereignty," a term that can mean various things ranging from having regulatory decisions made nationally instead of by Silicon Valley, and support for protectionist trade policies, to segmenting the Internet by blocking communications with other countries. In addition, national regulations to promote online competition, enhance cybersecurity, curtail hate speech, and protect citizens' data privacy might well vary significantly around the globe and even in geopolitical regions where there might otherwise be commonality. A rise in citizens' mistrust of governments (especially in the area of surveillance) might lead to a corresponding growth in the use of encrypted communications (both to avoid government surveillance and in response to general privacy concerns). Moreover, individual countries or blocs of like-minded countries might impose (or continue to impose) substantive communications content requirements enabled by technological distinctions at national levels, including, for example, banning or discouraging end-to-end encryption (so as to permit government surveillance), or mandating a variety of governmental access to otherwise encrypted communications (perhaps through required turnover of encryption keys to authorities or insisting on the use of specified encryption schemes).

Addressing the Challenges Posed by Encryption Requires Technical Talent That Is in Short Supply

It is a truism in the technology industry that cybersecurity talent is in short supply. Both government and industry are pursuing initiatives aimed at enlarging the pool of cybersecurity talent and filling hundreds of thousands or millions of openings.[17,18] The initiatives that make headlines are broad, focusing on roles ranging from system administrators to incident response personnel to software developers. But the need for talent also applies to the smaller and more highly trained populations of engineers who create secure software and systems and the researchers who develop and analyze new encryption algorithms and protocols. Addressing this need for talent will be a major challenge for the United States in the decades to come. While the Internet has enabled broader dissemination of knowledge about cryptography and security, the high-level research training and careers required to understand and contribute to the state of the art require real investment in research infrastructure from government and industry.

---

[17] D. Santos, 2021, "National Initiative for Cybersecurity Education (NICE)," National Institute of Standards and Technology, December 3, https://www.nist.gov/itl/applied-cybersecurity/nice.

[18] A. Counts, 2021, "Microsoft Wants 250,000 More Workers in Cybersecurity," Protocol—The People, Power and Politics of Tech, October 29, https://www.protocol.com/bulletins/microsoft-cybersecurity-talent-shortage.

Key Findings

FINDING 4.1: Most of the current public scientific expertise in algorithm design, cryptanalysis, and other areas of applied cryptography is outside the United States, largely in Europe. In contrast, within the United States, cryptography is taught as an area of theoretical computer science. The specific areas of expertise necessary to guide and facilitate the transition to post-quantum cryptography are relatively new and will require a more robust educational pipeline to train new talent.[19] Public research investment, through the National Science Foundation and other organizations, would encourage this process, while strict U.S. export control regulations have historically discouraged talent from locating in the United States.

FINDING 4.10: The United States needs far more data security expertise than is currently available, and these needs are growing substantially. The failure to meet these needs could have significant and widespread ramifications both for national security and the private sector. All software developers and computer scientists require basic competence in computer security. In addition, a growing number of people will require deep expertise in security. The required skills are not easy to teach, as students need both security-focused knowledge and a deep technical knowledge across multiple subjects and layers of abstraction. If the U.S. educational system does not meet these needs, or if the United States becomes a less attractive destination for students, researchers, and entrepreneurs born in other countries, the shortage will be much worse. Technological changes may rapidly increase demand for rare skills or may reduce demand by enabling tasks that currently require exceptionally skilled individuals to be performed by a broader range of people.

A Mathematical Breakthrough Could Threaten Current Encryption Algorithms

As discussed above, much of the current concern about the future of encryption has been motivated by the potential for development of working, large-scale quantum computers. The findings in this report discuss approaches to detecting progress toward an adversary's use of quantum computers. However, a disruptive breakthrough in mathematics that would improve the performance of conventional computers on specific problems relevant to decryption (such as factoring and discrete logarithms) would be much less costly and visible than the construction of a quantum computer and could provide significant offensive advantage.

Key Findings

FINDING 4.2: An improvement in asymmetric cryptanalysis algorithms could have a significant effect on the security of public key encryption algorithms that are in wide use today. Such an improvement would enable more efficient attacks on encrypted information using conventional computers rather than requiring the construction of a quantum computer. Furthermore, it could potentially be exploited in secret and with little or no advance notice.

FINDING 4.14: A new classical cryptanalysis algorithm or quantum computing development could result in rushed and disorganized efforts to replace widely used public key algorithms or other cryptographic standards. Such a breakthrough would require mitigation efforts that would be more complex than fixing typical software bugs, such as the coordinated deployment of major protocol updates across implementations and services.

The Lead of the Intelligence Community in Encryption Is Diminishing

The United States, and U.S. intelligence, have long been leaders in encryption and related areas, relative to foreign adversaries. As a result, in the absence of significantly increasing technical challenges that threatened to

---

[19] To understand the cryptographic landscape, one must receive a Ph.D. in cryptography with at least 3–5 years of highly specialized training in graduate school. Even though the information is freely available on the Internet, the sheer volume of information and high degree of specialization means that without hands-on advising, it is nearly impossible to learn the skillset necessary to become proficient in cryptography.

thwart its mission, the Intelligence Community had the luxury of continuing its relative superiority in this area. But the days of this superiority appear to be drawing to an end.

Key Finding

FINDING 6.1: With more adversary nations (especially China) seeking and making advances in encryption and as academic researchers (especially in Europe) continue to invest in cryptography and advance the theory and practice of encryption, the advantage that the Intelligence Community enjoyed in this area will diminish if not disappear.

Computing on Encrypted Data Has the Potential to Improve Security and Privacy for Individuals and Organizations

Historically, information had to be decrypted before it could be processed on computer systems. While decrypted, the information was a potential target for cyberattacks and subject to potential misuse by authorized system users. In recent years, there have been significant advances in algorithms that enable some kinds of processing of information without requiring that the information be decrypted. Computation on encrypted data at scale has the potential to enhance the sharing of intelligence products and enable intelligence activities that better protect individuals' privacy.

Key Finding

FINDING 2.4: The research community continues to make improvements in the technology of computation on encrypted data. Such improvements can be expected to enable new ways of securely sharing both government and private-sector information.

1

Introduction

Encryption is a process for making information unreadable.[1] The inverse process, making information that has been encrypted readable, is referred to as decryption. Encryption and decryption are facets of a broad scientific field referred to as cryptography. The statement of task for this study uses the term "encryption," but the scope of the study is the broader field of cryptography. (Historically, this field was referred to as "cryptology," but in current usage, the word "cryptography" is more common. This is the term the committee will use in this report.) This report sometimes uses the terms encryption and cryptography interchangeably; where the difference between encryption and decryption are significant, those terms are used in their technical sense.

Cryptography is a complex and specialized subject: Chapter 2 of this report introduces aspects of cryptography for readers who are not familiar with it, and the National Institute of Standards and Technology's (NIST's) glossary is a useful reference.[2]

For most of recorded history, encryption was an arcane process used primarily by governments, the military, the Roman Catholic Church, and a few commercial organizations that sought to protect their communications from disclosure.[3] Today, cryptography also enables authentication (verifying the identities of people, code, and the origins of transactions), and underlies the safe use of the Internet and computer systems by individuals and organizations worldwide.[4,5,6] Emerging cryptographic technologies offer capabilities such as the ability to process encrypted information without first decrypting it and distributing trust among multiple entities in a way that is

---

[1] D. Boneh and V. Shoup, 2020, "A Graduate Course in Applied Cryptography," Version 0.5, January, http://toc.cryptobook.us.

[2] National Institute of Standards and Technology (NIST), 2019, *Glossary of Key Information Security Terms*, NISTIR 7298 Revision 3, https://nvlpubs.nist.gov/nistpubs/ir/2019/NIST.IR.7298r3.pdf.

[3] D. Khan, 1967, *The Codebreakers: The Story of Secret Writing*, New York: Macmillan, pp. 106–107.

[4] D. Boneh and V. Shoup, 2020, "A Graduate Course in Applied Cryptography," Version 0.5, January, http://toc.cryptobook.us.

[5] J. Katz and Y. Lindell, 2021, *Introduction to Modern Cryptography*, Boca Raton, FL, Chapman & Hall/CRC Press, Taylor & Francis Group.

[6] O. Goldreich, 2004, *Foundations of Cryptography*, https://www.wisdom.weizmann.ac.il/~oded/foc.html.

resilient against malicious minorities of participants.[7,8,9,10,11] The emergence of distributed cryptocurrencies is built on these principles, but these technologies have far greater applicability.[12]

The U.S. Intelligence Community, like intelligence organizations worldwide, must use encryption to protect sensitive information from unauthorized disclosure or modification. It must also defeat the encryption of information that it collects as part of its mission. The Intelligence Community also has an interest in creating systems where even if there are malicious insiders, they cannot gain critical information or control. The protective use of encryption is referred to as "defensive" and the task of defeating encryption as "offensive." The defensive role of the U.S. Intelligence Community extends to setting standards and/or creating systems for the encryption of classified U.S. national security information and advising on the creation of standards for the encryption of unclassified government and private sector information.

The Office of the Director of National Intelligence (ODNI) requested that the National Academies of Sciences, Engineering, and Medicine establish a committee to identify potential scenarios for the balance between encryption and decryption over the next two decades and to assess the national security and intelligence implications of each scenario. The objective of the committee's effort is not to predict what developments will occur, but to identify the range of possible developments and their implications, and to provide the Intelligence Community with recommended ways of identifying which future scenarios are materializing so that U.S. intelligence and the U.S. government as a whole can respond to and take advantage of these changes.

STATEMENT OF TASK

The National Academies of Sciences, Engineering, and Medicine will convene an ad hoc committee to identify potential scenarios over the next 10 to 20 years for the balance between encryption and decryption (and other data and communications protection and exploitation capabilities). The committee will then assess the national security and intelligence implications of the scenarios it deems most relevant and significant, based on criteria it develops.

The committee will first identify plausible scenarios, and the technological drivers (and other major drivers as deemed relevant by the committee) behind these scenarios, and potential areas of technology surprise. It will consider such factors as likelihood, speed, difficulty of planning and response, and consequence, in order to advise on which scenarios are most worthy of attention. The committee will also consider implications for applications of encryption such as cybersecurity, digital currency, cybercrime, surveillance, and covert communication.

The committee will then assess the national security, intelligence, and broad societal implications of each scenario determined by the committee to be most worthy of attention; identify and assess options for responding to these scenarios; and assess the implications for future Intelligence Community investments. In doing so, it will consider actions common across all scenarios, scenario-dependent actions, and technology developments that the Intelligence Community should monitor in order to narrow the range of possible scenarios in the future. It will also consider how other governments might act in each of the scenarios, and the implications of those actions for the United States. This project will produce a peer-reviewed consensus report.

---

[7] O. Goldreich, S. Micali, and A. Wigderson, 2019, "How to Play Any Mental Game, or a Completeness Theorem for Protocols with Honest Majority," in *Providing Sound Foundations for Cryptography: On the Work of Shafi Goldwasser and Silvio Micali*, https://doi.org/10.1145/3335741.3335755.

[8] M. Ben-Or and A. Wigderson, 1988, "Completeness Theorems for Non-Cryptographic Fault-Tolerant Distributed Computation," *Proceedings of the Twentieth Annual ACM Symposium on Theory of Computing—STOC '88*, https://doi.org/10.1145/62212.62213.

[9] D. Chaum, C. Crépeau, and I. Damgard, 1988, "Multiparty Unconditionally Secure Protocols," *Proceedings of the Twentieth Annual ACM Symposium on Theory of Computing—STOC '88*, https://doi.org/10.1145/62212.62214.

[10] R. Cramer, 2015, *Secure Multiparty Computation and Secret Sharing*, Cambridge, UK: Cambridge University Press.

[11] C. Gentry, 2009, "Fully Homomorphic Encryption Using Ideal Lattices" in *Proceedings of the 41st Annual ACM Symposium on Theory of Computing—STOC '09*, https://doi.org/10.1145/1536414.1536440.

[12] J. Garay, A. Kiayias, and N. Leonardos, 2017, "The Bitcoin Backbone Protocol with Chains of Variable Difficulty," pp. 291–323 in *Advances in Cryptology—CRYPTO 2017*, https://doi.org/10.1007/978-3-319-63688-7_10.

THE COMMITTEE'S APPROACH AND PROCESS

The committee's work was conducted at the unclassified level and dealt only with cryptographic systems described by public sources. However, because governments and national security organizations worldwide are major users of commercial and public systems (including their security mechanisms), the results of this work apply to the challenges facing the Intelligence Community.

As a result of limitations on travel and face-to-face meetings resulting from the COVID-19 pandemic, the committee held short, frequent virtual meetings rather than the longer face-to-face meetings that would be typical for a National Academies' committee. The committee met virtually 27 times during the period between late August 2020 and September 2021.

The committee began its deliberations with a series of meetings aimed at reviewing the committee's task, introducing the committee members and their areas of expertise, and identifying issues that merited in-depth investigation. Following these meetings, the committee held eight meetings with experts and stakeholders who were invited to present their views on aspects of the future of encryption. Once the briefings were complete (in early 2021), the committee began the process of identifying scenarios and their implications. The committee's sponsor from the Office of the Director of National Intelligence met with the committee on two occasions, once during the early stakeholder briefings and later as the committee was refining its approach to identifying scenarios of interest. The full list of speakers and their topics is included in Appendix B.

To identify scenarios of interest to the Intelligence Community, the committee used a mix of approaches based on the work of futures and strategic foresight firms and academic researchers. The committee first identified "drivers" whose possible future states are important to the future of encryption and then used combinations of the extreme states of those drivers to define potential scenarios. The committee then selected the potential scenarios that are most informative and cover the broadest range of realistic futures and explored their possible implications for and impacts on the Intelligence Community.

This report goes into some depth in describing the drivers that determine the potential scenarios. As the committee explored and reviewed those drivers, it determined that some situations and trends in technology and policy have clear implications and impact no matter which direction the future takes. The committee documented those implications in the report's findings. The implications of and actions in anticipation of and response to the selected scenarios are included in the discussion of each scenario.

TRENDS AND MOTIVATIONS

The committee's work is well justified by the importance of encryption to the Intelligence Community and by the scope and variety of encryption-related changes in government policies and the commercial marketplace. However, one technical issue stands out as a particular motivator: the potential development of quantum computers that would rely on phenomena of quantum mechanics to perform computation in a fundamentally different way from the computers that have been built since the 1940s. A sufficiently large-scale, fault-tolerant quantum computer could be programmed to defeat almost all of the asymmetric[13] (i.e., public key) encryption and digital signature systems in current use on the Internet.[14] Specifically, the public-key encryption and digital signature algorithms in common use on the Internet today are based on problems like factoring that would be defeated by a quantum computer.[15] Public-key encryption schemes based on other mathematical hardness assumptions (such as finding

---

[13] Unlike symmetric encryption systems whose origins go back millennia, asymmetric encryption systems apply a public key that can be shared widely to encrypt information, and a separate private key, related to the public key by a hard-to-solve mathematical problem, to decrypt information. Most websites and services on the Internet rely on asymmetric encryption to achieve user authentication and data protection. A loss of public key cryptographic solutions would be difficult to replace, threatening the viability of a ubiquitous Internet.

[14] National Academies of Sciences, Engineering, and Medicine, 2019, *Quantum Computing: Progress and Prospects*, Washington, DC: The National Academies Press, https://doi.org/10.17226/25196.

[15] P.W. Shor, 1997, Polynomial-time algorithms for prime factorization and discrete logarithms on a quantum computer," *SIAM Journal on Computing* 26(5):1484–1509, https://doi.org/10.1137/s0097539795293172.

a shortest vector in a high-dimensional lattice) are expected to be standardized and deployed within the decade, and have so far resisted efficient attacks even if quantum computers are built at scale.[16,17]

Researchers, governments, and industry recognize the potential impact of quantum computers, and work is under way to identify, standardize, and implement "post-quantum" asymmetric encryption (and related digital signature) systems.[18] However, the transition from current public-key systems to post-quantum systems will require the modification of a vast amount of software and the replacement of some hardware that is fundamental to the operation of Internet-connected computer systems. The potential impact of quantum computers and the implications of the transition to post-quantum cryptographic systems were major subjects of the committee's work.

In addition to the potential impact of quantum computers, the committee addressed the impact of other trends in technology, policy, and society on the encryption issues that will face the Intelligence Community over the next one to two decades. The committee found that the impact of poor software and system design and implementation practices was a more significant threat to defense and a more significant opportunity for offense than the potential development of a quantum computer. Such poor practices are the norm today and offer attackers pervasive weaknesses to exploit. While some efforts are under way to improve design and implementation practices, they do not appear sufficient. This report includes findings on these issues and ways of addressing them.

OVERVIEW OF THE REPORT

Because cryptography is a technical subject, and an arcane one, the body of the report begins in Chapter 2 with an introduction to encryption and decryption (and more broadly, cryptography). Chapter 3 details the process that the committee used to identify potential scenarios for the future of encryption and its approach to selecting the scenarios that would be explored in depth. Chapter 4 provides both summaries and detailed descriptions of each of the three drivers that the committee determined to be critical in setting the path for the future of encryption. Chapter 5 summarizes the range of potential scenarios that the drivers define, presents the committee's rationale for selecting a specific subset of those scenarios, and describes the scenarios and their implications. Chapter 6 presents a brief discussion of the implications of the drivers and scenarios for the Intelligence Community.

The committee's findings are included in Chapters 2, 4, 5, and 6 in the context of the specific issues that motivate them. In addition, Chapter 7 summarizes the findings from the entire report.

---

[16] Z. Brakerski and V. Vaikuntanathan, 2014, Efficient fully homomorphic encryption from (standard) LWE, *SIAM Journal on Computing* 43(2):831–871, https://doi.org/10.1137/120868669.

[17] D. Micciancio and S. Goldwasser, 2013, *Complexity of Lattice Problems: A Cryptographic Perspective*, New York, Springer Science+Business Media.

[18] NIST, "Post-Quantum Cryptography," Computer Security Resource Center, Information Technology Laboratory, https://csrc.nist.gov/projects/post-quantum-cryptography, accessed October 12, 2021.

2

Introduction to Encryption

Modern cryptography uses techniques from mathematics to accomplish a variety of goals, including keeping data secret, authenticating data, and detecting dishonest behavior. Some techniques are in widespread use today; others are emerging. Each will be discussed in turn.[1,2,3]

CRYPTOGRAPHIC ALGORITHMS IN USE TODAY

This chapter begins by introducing the kinds of cryptographic algorithms most commonly used today. Historically, encryption was developed to ensure the confidentiality of data. Further cryptographic techniques in common use can be used to ensure the integrity or authenticity of data.

Private-Key Encryption

The simplest form of encryption is called *private-key encryption* or *symmetric encryption*.[4] To keep information (called plaintext) secret, the sender encrypts it by applying an algorithm to the plaintext and key to obtain ciphertext. The recipient can apply a second algorithm to the matching decryption key and ciphertext to decrypt the ciphertext and recover the original plaintext, but without it the ciphertext reveals no meaningful information about the plaintext. Figure 2.1 illustrates the operation of private-key encryption.

The most commonly used private-key encryption algorithm today is the Advanced Encryption Standard (AES) algorithm, which was chosen by the National Institute of Standards and Technology (NIST) in 2001 through a

---

[1] D. Boneh and V. Shoup, 2020, "A Graduate Course in Applied Cryptography," Version 0.5, January, http://toc.cryptobook.us.
[2] J. Katz and Y. Lindell, 2021, *Introduction to Modern Cryptography*, Boca Raton, FL: Chapman & Hall/CRC Press, Taylor & Francis Group.
[3] O. Goldreich, 2004, *Foundations of Cryptography*, https://www.wisdom.weizmann.ac.il/~oded/foc.html.
[4] O. Goldreich, S. Goldwasser, and S. Micali, 1986, How to construct random functions, *Journal of the ACM* 33(4):792–807, https://doi.org/10.1145/6490.6503.

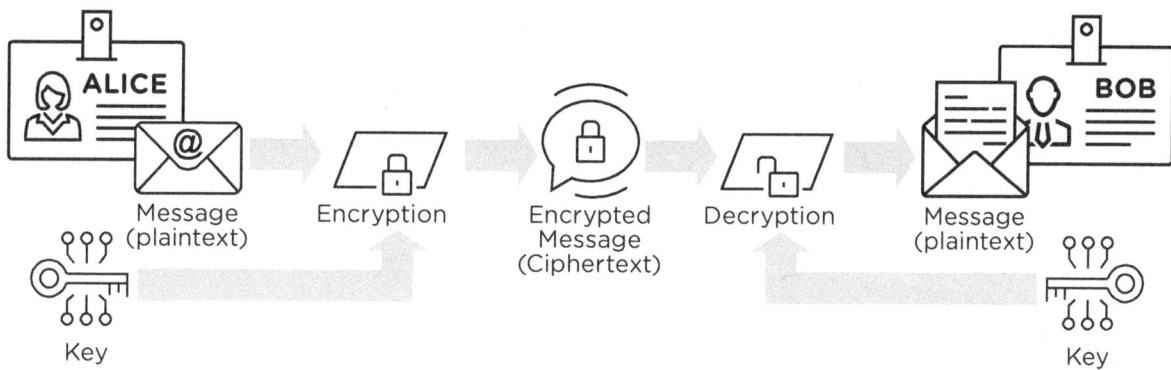

FIGURE 2.1 Private-key encryption.

public competition that solicited designs from cryptographers around the world.[5] It is fast enough that with a little hardware support, it is unlikely to be a bottleneck in a computer system.[6]

Symmetric encryption is widely used today to protect the confidentiality of data both "at rest"—that is, in storage—and "in transit"—that is, in communications. For example, a disk encryption system uses symmetric encryption to encrypt a user's data before writing it to disk and uses the same key to decrypt the data after reading from disk.[7] Security of the system resides in the security of the key:[8] an adversary who finds the key can also perform decryption (to steal the data) or encryption (to tamper with the data), so implementations use passwords, security hardware, or other measures to help protect keys.[9,10,11]

Public-Key Encryption and Key Distribution

With private-key or symmetric encryption, the encryption and decryption keys are the same, but with *public-key* or *asymmetric encryption*,[12,13] they are different and knowing the encryption key does not enable decryption. This means that if a message recipient publishes the encryption key and keeps its matching decryption key secret, anyone can send messages that only the recipient can read. Figure 2.2 illustrates the operation of public-key encryption.

Public-key encryption is used in practice to solve the (private) key distribution problem—that is, how can two parties agree on a shared symmetric encryption key if they have never communicated before, and are

---

[5] J. Daemen and V. Rijmen, 1999, "AES Proposal: Rijndael," National Institute of Standards and Technology, March 9, https://csrc.nist.gov/csrc/media/projects/cryptographic-standards-and-guidelines/documents/aes-development/rijndael-ammended.pdf.

[6] Intel, "Intel® Data Protection Technology with AES-NI and Secure Key," https://www.intel.com/content/www/us/en/architecture-and-technology/advanced-encryption-standard-aes/data-protection-aes-general-technology.html, accessed October 5, 2021.

[7] Dansimp, "Bitlocker (Windows 10)—Windows Security" (Windows 10)—Windows security | Microsoft Docs, https://docs.microsoft.com/en-us/windows/security/information-protection/bitlocker/bitlocker-overview, accessed October 5, 2021.

[8] "The concept that in a sound cryptosystem, the security of information depends solely on the key, is referred to as Kerchoff's principle." See D. Khan, 1967, *The Codebreakers: The Story of Secret Writing*, New York: Macmillan, p. 235.

[9] D. Boneh, S. Halevi, M. Hamburg, and R. Ostrovsky, 2008, "Circular-Secure Encryption from Decision Diffie-Hellman," pp. 108–125 in *Lecture Notes in Computer Science*, https://doi.org/10.1007/978-3-540-85174-5_7.

[10] J. Katz, R. Ostrovsky, and M. Yung, 2001, "Efficient Password-Authenticated Key Exchange Using Human-Memorable Passwords," pp. 475–494 in *Lecture Notes in Computer Science*, https://doi.org/10.1007/3-540-44987-6_29.

[11] Wikipedia, "Trusted Platform Module," Wikimedia Foundation, https://en.wikipedia.org/wiki/Trusted_Platform_Module, accessed October 14, 2021.

[12] S. Goldwasser and S. Micali, 1984, Probabilistic encryption, *Journal of Computer and System Sciences* 28(2):270–299, https://doi.org/10.1016/0022-0000(84)90070-9.

[13] D. Boneh and V. Shoup, 2020, "A Graduate Course in Applied Cryptography," Version 0.5, January, http://toc.cryptobook.us.

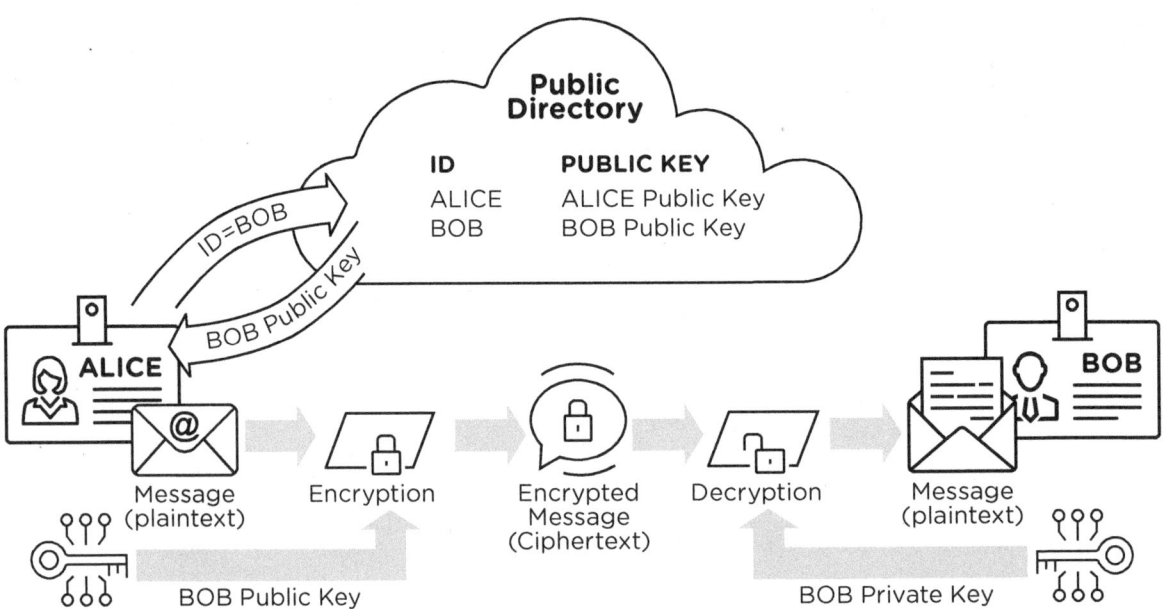

FIGURE 2.2 Public-key encryption.

communicating over an untrusted channel? To solve this problem using public-key cryptography, one party can choose a random secret symmetric key, encrypt it with the other party's public key, and send it to the other party. (Public-key encryption is much slower than symmetric encryption, so it is ordinarily used only to exchange private keys, and symmetric algorithms are used to protect the actual message data.[14])

In 1976, Diffie and Hellman published a paper in the open literature that built on Merkle's concept of public key cryptography[15] and introduced the technique for key agreement that is now known as Diffie-Hellman key exchange.[16] This algorithm is still used today in the form originally published, as well as a variant using elliptic curves (see below) that is somewhat more efficient. A widely used algorithm for public-key encryption is known as RSA, named for its creators Rivest, Shamir, and Adleman, who published the algorithm in 1977.[17]

To the non-technical person, public-key cryptography may seem like magic, as it depends on certain mathematical operations being easy to do but hard to undo. For example, the security of RSA depends on the fact that it is computationally fast to multiply two large prime integers but, as far is known, there is no computationally fast general algorithm to recover the two factors from the product with a classical computer. Elliptic curve cryptography, based on the algebraic structure of elliptic curves over finite fields, depends on the fact that it is easy to multiply a publicly known base point by a scalar value but as far as is known there is no computationally efficient algorithm to compute the corresponding scalar given just the base and target point.[18]

---

[14] Ellis, Cocks, and Williamson developed ideas behind public-key cryptography, which they called "non-secret encryption," in classified reports at GCHQ. This history has since been declassified. See J.H. Ellis, "The History of Non Secret Encryption," last modified August 28, 2016, https://www.cs.miami.edu/home/burt/manuscripts/crypto_for_intelligence/ellis.pdf.

[15] R. Merkle, "Secure Communications Over Insecure Channels," http://www.merkle.com/1974/PuzzlesAsPublished.pdf.

[16] W. Diffie and M. Hellman, 1976, New directions in cryptography, *IEEE Transactions on Information Theory* 22(6):644–654, https://doi.org/10.1109/tit.1976.1055638.

[17] R.L. Rivest, A. Shamir, and L. Adleman, 1978, A method for obtaining digital signatures and public-key cryptosystems, *Communications of the ACM* 21(2):120–126, https://doi.org/10.1145/359340.359342.

[18] D. Boneh and V. Shoup, 2020, "A Graduate Course in Applied Cryptography," Version 0.5, January, http://toc.cryptobook.us.

To give a concrete use case, public-key cryptography is widely used as part of secure communications protocols on the Internet. The first step of many encrypted communications protocols like transport layer security (TLS), which is used to encrypt Hypertext Transfer Protocol Secure (HTTPS) web traffic, or Internet protocol security (IPsec), which is often used for virtual private networks (VPNs), is for both parties to perform a public key-based key exchange so that both sides agree on a shared secret to use for authentication and symmetric encryption.[19,20]

Digital Signatures

Digital signature schemes use mathematical techniques similar to those used in public-key encryption schemes, but with the specific purpose of authenticating data.[21] A digital signature scheme allows anyone in possession of the public verification *key* to verify a digital signature of a message, but only someone in possession of the matching private *signing key* can sign the message. Figure 2.3 illustrates the operation of digital signature systems.

If a message originator publishes the verification key, anyone can check that the plaintext is the same as the message that was signed. If the signing key is kept secret, then only the entity with the private signing key could have generated a valid signature. Digital signature algorithms in common use today include the RSA digital signature algorithm and the Elliptic Curve Digital Signature Algorithm (ECDSA).[22,23]

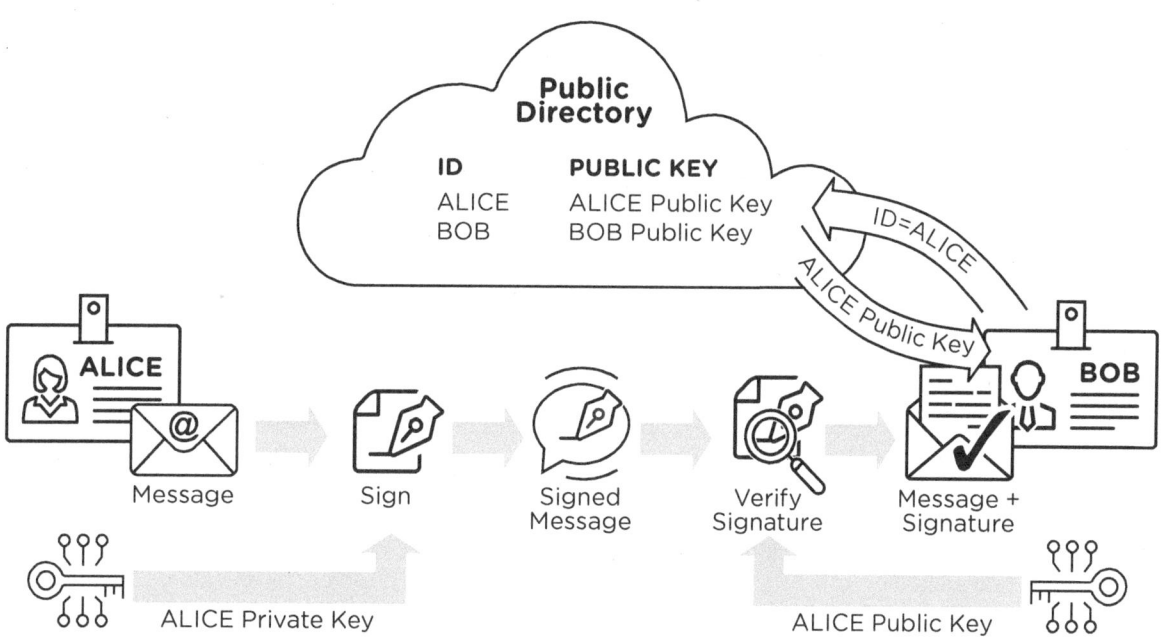

FIGURE 2.3 Digital signature systems.

---

[19] Wikipedia, "Transport Layer Security," Wikimedia Foundation, https://en.wikipedia.org/wiki/Transport_Layer_Security, accessed October 6, 2021.

[20] Wikipedia, "IPsec," Wikimedia Foundation, https://en.wikipedia.org/wiki/IPsec, accessed September 29, 2021.

[21] S. Goldwasser, S. Micali, and R.L. Rivest, 1988, A digital signature scheme secure against adaptive chosen-message attacks, *SIAM Journal on Computing* 17(2):281–308, https://doi.org/10.1137/0217017.

[22] R.L. Rivest, A. Shamir, and L. Adleman, 1978, A method for obtaining digital signatures and public-key cryptosystems, *Communications of the ACM* 21(2):120–126, https://doi.org/10.1145/359340.359342.

[23] D. Johnson, A. Menezes, and S. Vanstone, 2001, The Elliptic Curve Digital Signature Algorithm (ECDSA), *International Journal of Information Security* 1(1):36–63, https://doi.org/10.1007/s102070100002.

There are several major applications of public-key signature schemes. One is for software updates: the vendor signs the update, and then any client that knows the vendor's public key can verify that the bits it is about to install actually came from the vendor. Other applications of digital signatures include authenticating user logins, securing credit card payments, cryptocurrencies, signing documents, and as a part of security protocols like TLS to detect tampering with messages. (See Box 2.1 for an introduction to TLS.)

Another use of digital signatures is to create digital certificates. A certificate is a digitally signed message that binds together a public key and an entity's identity. For example, a certificate for a website includes the domain name (e.g., whitehouse.gov) and the web server's public key. The digital signature on a certificate may be generated by a third party called a certificate authority; the certificate authority's own verification key is usually distributed with an operating system or web browser. Often, there may be a chain of signed and trusted certificates and public keys that end in a key that is explicitly trusted. The security of a public key infrastructure using digital certificates also depends on policies and practices related to the issuance, storage, expiration, and revocation of credentials.[24,25]

Collision-Resistant Hash Functions

A *collision-resistant hash function* boils down an arbitrarily large source message to a small output value called a *hash* (perhaps 256 bits) in such a way that finding a different source message with the same hash (a collision) takes too much computational power to be practical.[26,27] Unlike encryption functions, cryptographic hash functions in use today are not parameterized by a key; instead, for a given hash function the same input will always result in the same output. The SHA-2 algorithm is a common hash algorithm in use today, and the SHA-3 algorithm family is becoming more popular.[28] The MD5 hash algorithm is a historically very popular algorithm that remains in common use today, despite being known to be insecure because researchers have demonstrated how to produce collisions.[29,30] (See the section on Algorithm Transitions later in this chapter for a further discussion of attacks leveraging the insecurity of MD5.)

Hash functions are an important ingredient in many cryptographic applications, including most uses of digital signatures. The short, fixed-length output size of a hash function is useful because public-key signatures are slow. Applying a hash function to the message first means that the digital signature algorithm can be applied to the much shorter hash value. Hash functions have many other common uses, including being used to verify file integrity and to construct blockchains for cryptocurrencies.[31]

User Authentication

The above cryptographic tools can help secure protocols for user authentication and minimize information exposure if the system is compromised. For example, storing cryptographic hashes of passwords (rather than the passwords themselves) can make it somewhat more difficult for attackers who compromise a server to determine user passwords. Hardware-based tokens or keycards can use digital signatures to implement a challenge-response authentication protocol.

Biometrics are a class of authentication techniques that use properties of individuals (e.g., fingerprints, palm prints, iris scans, facial recognition, etc.) to authenticate presence instead of (or in addition to) cryptographic

---

[24] P.C. Kocher, 1998, "On Certificate Revocation and Validation," pp. 172–177 in *Financial Cryptography*, https://doi.org/10.1007/bfb0055481.

[25] W. Aiello, S. Lodha, and R. Ostrovsky, 1998, "Fast Digital Identity Revocation," pp. 137–152 in *Advances in Cryptology—CRYPTO '98*, https://doi.org/10.1007/bfb0055725.

[26] D. Boneh and V. Shoup, 2020, "A Graduate Course in Applied Cryptography," Version 0.5, January, http://toc.cryptobook.us.

[27] Y. Ishai, E. Kushilevitz, and R. Ostrovsky, 2005, "Sufficient Conditions for Collision-Resistant Hashing," pp. 445–456 in *Theory of Cryptography*, https://doi.org/10.1007/978-3-540-30576-7_24.

[28] Wikipedia, "SHA-2," Wikimedia Foundation, https://en.wikipedia.org/wiki/SHA-2, accessed September 27, 2021.

[29] R. Rivest, 1992, "The MD5 Message-Digest Algorithm," https://doi.org/10.17487/rfc1321.

[30] X. Wang and H. Yu, 2005, "How to Break MD5 and Other Hash Functions," pp. 19–35 in *Lecture Notes in Computer Science*, https://doi.org/10.1007/11426639_2.

[31] C. Badertscher, U. Maurer, D. Tschudi, and V. Zikas, 2017, "Bitcoin as a Transaction Ledger: A Composable Treatment," pp. 324–356 in *Advances in Cryptology—CRYPTO 2017*, https://doi.org/10.1007/978-3-319-63688-7_11.

BOX 2.1
The SSL/TLS Protocol

For even a single network switch, router, or communications link, security audits are expensive, slow, and miss vulnerabilities. The Internet involves myriad components that are geographically dispersed, built by numerous vendors, and optimized for performance rather than security. The Internet also connects innumerable entities with opposing interests, so there is no party who is broadly trusted (much less sufficiently empowered or funded) to conduct mandatory comprehensive audits. As a result, no company or government will ever have any hope of gaining meaningful trust in the security of the Internet's communications channels. The Secure Sockets Layer (SSL)/TLS protocol[a,b] addresses this problem by allowing entities to communicate securely without having to trust (or even understand) the underlying network.

In the protocol's initial ("handshake") phase, the devices agree on a set of cryptographic algorithms they both support. They use the selected public key algorithm, typically elliptic curve Diffie-Hellman or RSA, to establish a set of shared symmetric keys. As used for typical web traffic, the server also supplies a certificate that specifies the server's public key and domain name so that a server authorized for "example.com" cannot, for example, masquerade as "irs.gov." The browser must verify that the certificate is valid, matches the domain in the URL, and is signed by an issuer whose public key is embedded in the browser.

Symmetric cryptography is much more efficient than public key cryptography and is used to protect the bidirectional exchange of payload data, such as the URL being requested and web page being returned. The transmitting party encrypts the data to be sent (for secrecy) and adds a message authentication code (MAC) (to detect tampering). The receiving party knows the same symmetric keys as the sender and can decrypt the data and recompute the MAC. If the computed MAC differs from the received value, the receiver treats this as a fatal error, because SSL/TLS assumes that lower-level network protocols (e.g., TCP) will detect and correct any non-malicious communication errors.

The SSL/TLS protocol assumes that both the client and server are trusted and they are implemented correctly. Implementation bugs can expose keys or traffic to attackers, so organizations make significant effort to secure their implementations. In addition, some trust is placed in the certificate authorities whose certificates are accepted by browsers, because they have the power to issue unauthorized certificates that would let an adversary impersonate a domain. To help make such abuses detectable, certificate authorities are required to publicly document[c] the certificates they issue so that domain owners can detect any improper certificate issuances.

SSL/TLS was designed to be able to evolve. For example, if a cryptographic algorithm is broken, it can be deprecated and will not be used if either the client or server implementations remove support. Likewise, new algorithms can be added without breaking backward compatibility as long as the communicating devices have at least one algorithm in common. Because of this flexibility, the process of adding quantum-resistant public key algorithms to SSL/TLS should be smoother than for many other protocols. Still, SSL/TLS's architecture does not meet the needs of all applications. For example, messaging and video conferencing systems often need to allow many participants, while SSL assumes a client-server model. Likewise, SSL/TLS uses symmetric MACs, which are not well suited to after-the-fact verification by third parties, so payment applications may apply digital signatures to each transaction. SSL/TLS does not protect against traffic analysis, so other protocols are better suited to applications that need to conceal information such as source and destination addresses or the sizes and timing of packets. As a result, while SSL/TLS will likely remain widely used, new cryptographic protocols will also be needed to meet the world's diverse and changing security needs.

---

[a] Network Working Group, 2008, "The Transport Layer Security (TLS) Protocol, Version 1.2," Request for Comments 5256, https://datatracker.ietf.org/doc/html/rfc5246.

[b] Internet for Engineering Task Force, 2018, "The Transport Layer Security (TLS) Protocol Version 1.3," Request for Comments 8446, https://datatracker.ietf.org/doc/html/rfc8446.

[c] MDN, "Certificate Transparency," last modified May 20, 2022, https://developer.mozilla.org/en-US/docs/Web/Security/Certificate_Transparency.

secrets known to the individual. Because it is difficult to memorize cryptographic keys, a common scenario for cryptographic key storage is to store the key on a device that uses a fingerprint reader or facial recognition to unlock access to the key.

Random Number Generators

Keys need to be unguessable, and one way to ensure this is to generate them at random. Computers can be equipped with hardware that allows them to collect environmental randomness by observing a physical phenomenon. For example, many central processing units (CPUs) today include hardware random number generators based on thermal noise. Approaches based on more exotic principles, such as radioactive decay or quantum optical sources, can also be theoretically sound, but are difficult or expensive to implement in practice. In recent years, quantum optical sources of randomness have become popular and may prove significantly faster than traditional techniques; however, they too are subject to failures such as producing less entropy than expected. Physical entropy sources may be slow or produce somewhat biased measurements (such as bits that do not have exactly 50 percent odds of being 0 versus 1), so modern computer systems use algorithms called pseudo-random number generators (PRNGs) that take as input an unpredictable starting string of bits (called a "seed") and produce a much larger "random-looking" sequence of bits that cannot be distinguished from truly random bits without knowing the seed.[32,33] These algorithms are a fundamental building block for modern cryptography, and many constructions use symmetric ciphers, hash functions, or asymmetric ciphers as building blocks.[34]

Box 2.2 defines blockchain and how it relates to cryptocurrencies.

BOX 2.2
Blockchains and Cryptocurrencies

A *blockchain* is a data structure that uses cryptographic techniques to grow and maintain an append-only list of *blocks* (records). Each block contains new data records along with a timestamp and a cryptographic hash of the previous block. Including hashes of prior blocks in every new block added means that modifications can be detected because any change to the data in a prior block will change every subsequent block. The process of adding new blocks to an existing blockchain can either be centralized or be performed in a decentralized manner by a group of parties that use a consensus protocol to reach agreement. Properly constructed, public blockchains can be visible, tamper-resistant, and resilient data structures for storing public information.

One application of blockchains is to record financial transactions for cryptocurrencies. In a cryptocurrency application, each block in the blockchain contains a list of new cryptocurrency transfers ("X currency units transfer from public key A to public key B" with a signature produced using private key A). In most common cryptocurrencies, participation is open and pseudonymous; participants are identified only by an identifier derived from a public key to which they hold the corresponding private key. The transaction completes when it is included in the common blockchain.

Participants in Bitcoin and most other cryptocurrencies are identified only by cryptographic keys, but transactions are public, so analysis of the blockchain can identify sources and destinations of cryptocurrency value. To enhance anonymity, cryptocurrencies such as Zcash leverage a cryptographic method known as zk-SNARKs (see discussion of Zero Knowledge below) to hide the keys involved in transactions, preventing third parties from tracing the flow of funds from one transaction to the next.

---

[32] M. Blum and S. Micali, 1984, How to generate cryptographically strong sequences of pseudorandom bits, *SIAM Journal on Computing* 13(4):850–864, https://doi.org/10.1137/0213053.

[33] O. Goldreich, S. Goldwasser, and S. Micali, 1986, How to construct random functions, *Journal of the ACM* 33(4):792–807, https://doi.org/10.1145/6490.6503.

[34] D. Boneh and V. Shoup, 2020, "A Graduate Course in Applied Cryptography," Version 0.5, January, http://toc.cryptobook.us.

CRYPTANALYSIS AND FOUNDATIONS OF TRUST IN CRYPTOGRAPHY

The following paragraphs turn to a discussion of how the cryptographic community establishes trust in cryptographic algorithms.

There are multiple ways that encrypted data might be compromised by an adversary. First, an adversary could steal the key—for example, by compromising a person or computer that knows it. No matter how strong a cipher is, if the key is compromised, then anyone who gets access to the decryption key can decrypt and recover the encrypted data. Hiding the data necessarily requires keeping decryption keys secret.

Second, the cryptographic algorithm might contain a weakness that makes it practical for an adversary to decode the data without stealing the key. For example, in World War II, the Army Signals Intelligence Service was able to break the Japanese cipher code referred to as Purple by analyzing encrypted messages.[35]

Modern ciphers are expected to withstand such attacks. Trust in cryptographic algorithms is built over time by community consensus about which algorithms resist all known cryptanalytic attacks. Over the past decades, cryptographers have rigorously defined various mathematical and security properties. When new algorithms and protocols are proposed, their designers often include statements about the intended security properties, as well as evidence such as proofs of equivalence to well-studied mathematical problems. Still, there is no absolute guarantee that any of the underlying mathematical problems cannot be solved efficiently.[36,37] As a result, confidence in systems involves public analysis of the designer's assumptions and claims, and widely used algorithms are continuously analyzed by the international community of cryptography researchers. In addition, there are typically vast safety margins between the believed strength of a cipher and what adversaries can plausibly muster. For vendors, there is an increasing expectation that implementations should be subjected to security audits and testing.

Over the past decades, NIST has overseen the creation of several widely respected and adopted cryptographic algorithm standards. The first such standard issued by NIST was the Data Encryption Standard (DES),[38] a cryptographic block cipher designed by NIST in consultation with the National Security Agency (NSA) and based on a proposal submitted by IBM. When NIST determined in 1997 that it was time to retire DES and replace it with a new block cipher, NIST held a competition to select a new algorithm to be the replacement Advanced Encryption Standard (AES). NIST's AES competition was an open, transparent, and public process, attracting 15 algorithm submissions from around the world. Over the next several years and after three rounds of public analysis, in 2000 NIST selected the Rijndael[39] submission as the winner and subsequently issued Federal Information Processing Standard (FIPS) 197, formally defining AES.

The process that NIST established and followed to select AES has become the benchmark for subsequent open and transparent algorithm competitions worldwide. NIST conducted an open competition to select the Secure Hash Algorithm-3 (SHA-3)[40] standard, receiving 51 candidate submissions in 2008 and eventually selecting the Keccak algorithm as the winner. Currently, NIST is conducting an open competition to select a set of post-quantum public-key cryptographic algorithms; more than 80 proposals from around the world were submitted to NIST's post-quantum cryptography (PQC) standardization process[41] in November 2017, and NIST is expected to select multiple winning algorithms starting in mid-2022.

---

[35] J.R. Leutze, 1983, "Ronald Lewin. *The American Magic: Codes, Ciphers, and the Defeat of Japan*," *American Historical Review*, https://doi.org/10.1086/ahr/88.1.211.

[36] A prominent exception is the one-time pad, which Shannon proved was information-theoretically secure in a landmark paper. Unfortunately, the difficulty of properly generating and distributing key material renders the one-time pad difficult to impossible to use in practice, and there have been numerous intelligence failures owing to these usability issues.

[37] Such a proof would imply that P != NP, and would mean resolving one of the most famous unsolved problems in mathematics and computer science.

[38] NIST, 1977, "Data Encryption Standard," Federal Information Processing Standards Publication 46, January 25.

[39] J. Daemen and V. Rijmen, 2005, "Rijndael/AES," in *Encyclopedia of Cryptography and Security* (H.C.A. van Tilborg, ed.), https://doi.org/10.1007/0-387-23483-7_358.

[40] G. Bertoni, J. Daemen, M. Peeters, and G. Van Assche, 2013, "Keccak," pp. 313–314 in *Advances in Cryptology—EUROCRYPT 2013*, https://doi.org/10.1007/978-3-642-38348-9_19.

[41] NIST, "Post-Quantum Cryptography Standardization—Post-Quantum Cryptography," Computer Security Resource Center, Information Technology Laboratory, https://csrc.nist.gov/projects/post-quantum-cryptography/post-quantum-cryptography-standardization.

In addition to open competitions, sometimes NIST has partnered with other agencies or standards organizations to endorse cryptographic standards not produced by an open, transparent, and public process. For example, the 1995 SHA-1 hash algorithm standard and the 2000 SHA-2 family of hash algorithms were all developed initially by NSA and subsequently standardized by NIST. Another noteworthy standard not resulting from an open competition was the 2015 Dual Elliptic Curve Deterministic Random Bit Generator standard (DUAL-EC). Developed initially as part of an American National Standards Institute (ANSI) standards activity in which NIST participated, the DUAL-EC standard was recognized early on as having an unusual design structure with specific constants that had been defined by NSA. In 2007, Shumow and Ferguson[42] showed that if the constants were constructed in a particular way, knowledge of that secret construction would allow one to predict outputs of the random number generator, thus compromising its security. Although there was no allegation in 2007 that the constants in the standard were deliberately constructed in this way, reporting following the 2013 Snowden disclosures alleged exactly this. The allegation (never confirmed or refuted by NSA) eventually led to the DUAL_EC_DRBG standard being retracted by NIST and removed from implementations around the world.[43]

The DUAL_EC incident weakened confidence in NIST's process for creating cryptographic standards. In response, NIST sponsored an external review of its processes by a committee appointed by its Visiting Committee on Advanced Technology (VCAT) and accepted the recommendations of the review.[44] The recommendations included increasing the openness and transparency of NIST's processes for developing cryptographic standards, increasing NIST's independent cryptographic capabilities, and clarifying the relationship between NIST and NSA with regard to the development of cryptographic standards. However, this incident led to lingering concerns about other NIST standards, particularly those related to elliptic curve cryptography. These concerns led some other standards bodies, most notably the Internet Engineering Task Force, to adopt additional elliptic curves in addition to the NIST-standardized curves. These additional curves did not go through a NIST competition or similar NIST standardization process, yet the end result was that NIST was forced to include them in a revision to NIST's elliptic curve cryptography standard.

Thus, standards setting processes can be seen as dynamic. When new threats emerge, needs change, or scientific understanding improves, stronger algorithms are devised, studied, standardized, and deployed. Algorithms that do not come from an open process such as the one that NIST has frequently used, or that are not subject to years of public scrutiny, tend to be regarded as less likely to be secure. This skepticism has often been for good reason.[45,46] See for instance the example detailed in the article by Dunkelman, Keller, and Shamir.[47] Establishment of trust may involve different entities doing separate reviews. For example, multiple governments may separately review a cipher and conclude that it is acceptable for their own sensitive information.

Classical Cryptanalysis

If an adversary can correctly guess the value of a decryption key, the guessed key will decrypt the message, and security is compromised. Thus, to stop attacks that involve guessing secret keys, the number of possible

---

[42] D. Shumow and N. Ferguson, 2007, "On the Possibility of a Back Door in the NIST SP800-90 Dual Ec Prng," http://rump2007.cr.yp.to/15-shumow.pdf.

[43] D.J. Bernstein, T. Lange, and R. Niederhagen, 2016, "Dual EC: A Standardized Back Door," pp. 256–281 in *The New Codebreakers: Essays Dedicated to David Kahn on the Occasion of His 85th Birthday* (P.Y.A. Ryan, D. Naccache, and J.-J. Naccache, eds.), Berlin: Springer, https://eprint.iacr.org/2015/767.

[44] NIST, 2014, "VCAT Report on NIST Cryptographic Standards and Guidelines Development Process," July, https://www.nist.gov/system/files/documents/2017/05/09/VCAT-Report-on-NIST-Cryptographic-Standards-and-Guidelines-Process.pdf.

[45] E. Biham and L. Neumann, 2018, "Breaking the Bluetooth Pairing: Fixed Coordinate Invalid Curve Attack," updated July 25, http://www.cs.technion.ac.il/~biham/BT.

[46] A. Biryukov, A. Shamir, and D. Wagner, 2001, "Real Time Cryptanalysis of A5/1 on a PC—Springer," Chapter 10 in *Fast Software Encryption* (G. Goos, J. Hartmanis, J. van Leeuwen, and B. Schneier, eds.), FSE 2000, Lecture Notes in Computer Science, Volume 1978, Berlin, Heidelberg: Springer, https://doi.org/10.1007/3-540-44706-7_1.

[47] O. Dunkelman, N. Keller, and A. Shamir, 2010, "A Practical-Time Attack on the A5/3 Cryptosystem Used in Third Generation GSM Telephony," https://eprint.iacr.org/2010/013; European Agency for Cybersecurity, 2017, "An Overview of the Wi-Fi WPA2 Vulnerability," https://www.enisa.europa.eu/publications/info-notes/an-overview-of-the-wi-fi-wpa2-vulnerability.

cryptographic keys must be much larger than any plausible adversary could try. Current industry recommendations usually suggest at least 128 bits of security, so that the estimated running time of a brute-force attack (e.g., trying all possible 128-bit keys) would be 2 to the power of 128 (i.e., 2 * 2 * 2 * ...* 2, or 340282366920938463 463374607431768211456).[48] To get a sense of the size of this number, an attacker with a billion computers, each trying a billion keys per second, would need 10.8 trillion years to try all values. When there are attacks against a cryptographic algorithm that can figure out the key or recover information about plaintexts faster than trying all possibilities, the "strength" of the algorithm is measured by the running time of the best-known attack.

For symmetric encryption algorithms such as AES that are believed to be secure, the best currently known attacks using classical computers are no faster than trying all possible keys. For collision-resistant hash functions such as SHA-3, the best-known attacks are exponential in half the output length. For example, SHA-3 with 256-bit output is believed to provide equivalent strength to AES-128.

In contrast, the known methods for constructing efficient public-key encryption and digital signature algorithms take advantage of sophisticated mathematical structures to enable the proper functioning of public-key cryptography. These mathematical operations make public-key encryption less efficient than the symmetric algorithms in common use (and thus make public-key encryption usable for key distribution but inefficient for protecting large messages). This mathematical structure means that some of these algorithms are vulnerable to attacks that are much more efficient than brute force, and thus require longer keys to achieve security levels equivalent to a desired bit strength.[49] For a 2048-bit RSA key, the best-known classical attack is estimated to be equivalent to brute-forcing a 112-bit symmetric key, and 3072-bit RSA is considered to provide 128-bit security. The security of elliptic curve–based public key systems scales differently, and a 256-bit elliptic curve key is currently considered to provide 128 bits of security against a classical computer.

Quantum Computing and Quantum Cryptanalysis

Since the 1990s, it has been known that quantum computers could theoretically break some (but not all) public-key cryptographic algorithms far faster than classical computers could do so. Sufficiently large quantum computers cannot yet be realized in practice, but if they are eventually developed then future adversaries might use them to decrypt communications that have previously been collected and archived.

Quantum computers use quantum mechanical effects in ways that are fundamentally different from classical computers. Where classical computers store information as binary digits—*bits*—that may be either 0 or 1 in value, quantum computers store information in quantum bits—*qubits*—that are two-state quantum mechanical systems that contain superpositions of both 0 and 1 values at the same time. Quantum computers use qubits and their quantum-mechanical properties like superposition and entanglement in order to perform some computations much faster than would be possible with a conventional computer.

Algorithms for quantum computers are described in terms of idealized "logical" qubits, but efforts to build quantum computers in the real world produce physical qubits that degrade quickly and exhibit errors when measured. At present, quantum computers consisting of a small number of physical qubits (fewer than 100) have been realized. (A different approach, taken by the company D-Wave Systems, is to build quantum computers that perform a process called "quantum annealing." These computers can reach much larger numbers of qubits but cannot perform the operations needed to break RSA and other public key cryptosystems.) It is understood in theory how to use error correction techniques to combine many imperfect physical qubits into a single stable, logical qubit, but

---

[48] For classified uses up to TOP SECRET, NSA's Commercial National Security Algorithms (CNSA) Suite specifies algorithms with at least 192 bits of security. See NSA, 2015, "Commercial National Security Algorithm Suite," reviewed August 19, https://apps.nsa.gov/iaarchive/programs/iad-initiatives/cnsa-suite.cfm.

[49] The best public attacks against RSA moduli have been performed using the general Number Field Sieve (NFS). The RSA-768 challenge number, a 768-bit, 232-digit RSA modulus, was successfully factored in 2010 using NFS (see K. Aoki, J. Franke, A.K. Lenstra, E. Thomé, J.W. Bos, P. Gaudry, A. Kruppa, et al., 2010, "Factorization of a 768-bit RSA Modulus," Version 1.4, February 18, https://eprint.iacr.org/2010/006.pdf). The current record for factoring was reported in February 2020 and is held by Boudot, Gaudry, Guillevic, Heninger, Thome, and Zimmermann (see P. Zimmermann, 2020, "[Cado-nfs-discuss] Factorization of RSA-250," February 28, https://sympa.inria.fr/sympa/arc/cado-nfs/2020-02/msg00001.html), who factored the RSA-250 challenge number with NFS using approximately 2,700 core-years in total CPU time.

figuring out how to do so in practice and then scaling many logical qubits into a fault-tolerant quantum computer are areas of active research in the field of quantum computing.

Secret-key encryption and hash functions do not seem to have a mathematical structure that is amenable to attack by quantum computers beyond generic speedups for brute-force search. In particular, the best attacks currently known reduce the effective size of symmetric encryption keys by no more than a factor of 2. Equivalently, a brute-force key search on a quantum computer takes the square root of the running time it would take on a classical computer. This speedup is owing to Grover's algorithm (1996), a quantum search algorithm.[50] In concrete terms, to find an AES-256 key, a quantum computer would need to perform at least as many operations as a classical computer would need to break AES-128—a huge number. (Quantum computers are not believed to provide any improvement against collision-resistant hash functions like SHA-3.) Thus, if large-scale fault-tolerant quantum computers are ever realized, symmetric key sizes must be doubled to maintain the same strength against attacks. In practice, this is reasonably straightforward, as symmetric algorithms such as AES that support 256-bit keys have been available for many years.

In contrast, all public-key schemes in common use today have a mathematical structure that allows an efficient attack given a sufficiently powerful quantum computer. The strength of the RSA cryptosystem is based on the difficulty of factoring, and an efficient quantum algorithm for factoring large numbers—Shor's algorithm—was developed in 1994. Variants of Shor's algorithm can break Diffie-Hellman key agreement and the elliptic curve-based algorithms Elliptic Curve Diffie-Hellman (ECDH) and ECDSA. Although the number of qubits needed depends on the public-key algorithm and size of the public key, at about 2400 logical qubits quantum computers would start to break modern public-key implementations.

The threat of future cryptographically relevant quantum computers has already spurred preparations for migrating to new, quantum-resistant public-key algorithms. Currently, NIST is running a multi-year open competition to produce recommended algorithms for key exchange, public key encryption, and digital signatures that are resistant to attacks by quantum computers.[51] The new algorithms need to have a mathematical structure that is entirely different from common algorithms in use today in order to escape known attacks by quantum computers. Such algorithms are commonly called "post-quantum" or "quantum-resistant" cryptography.[52]

The new cryptographic algorithms under consideration are based on a few different families of mathematical hardness assumptions. One of these new candidate assumptions, which underlies several of the encryption and signature algorithms under consideration as of this writing, is that finding a shortest non-zero vector in a high-dimensional lattice is difficult even for quantum computers.[53] Other candidate encryption and digital signature algorithms include algorithms based on the difficulty of decoding certain error-correcting codes, solving systems of multivariate equations, and schemes based on non-interactive, zero-knowledge protocols (see below).

For digital signatures, some quantum-resistant approaches, referred to as stateful digital signature, are constructed using only collision-resistant hash functions.[54,55] These are very efficient, and do not rely on the relatively new and more structured mathematical assumptions of the other quantum-resistant public key algorithms.

FINDING 2.1: Stateful digital signatures based on hash functions are practical today and will remain secure even if large-scale quantum computers are practical or if new number theoretic attacks are developed that

---

[50] L.K. Grover, 1996, "A Fast Quantum Mechanical Algorithm for Database Search," *Proceedings of the Twenty-Eighth Annual ACM Symposium on Theory of Computing—STOC '96*, https://doi.org/10.1145/237814.237866.

[51] NIST, "Post-Quantum Cryptography Standardization—Post-Quantum Cryptography," Computer Security Resource Center, Information Technology Laboratory, https://csrc.nist.gov/projects/post-quantum-cryptography/post-quantum-cryptography-standardization.

[52] Post-quantum encryption algorithms are so named because they are intended to resist cryptanalysis by a quantum computer. They run on a conventional computer (not a quantum computer) and are unrelated to quantum key distribution or quantum networks (discussed below).

[53] D. Micciancio and S. Goldwasser, 2002, "Cryptographic Functions," pp. 143–194 in *Complexity of Lattice Problems*, https://doi.org/10.1007/978-1-4615-0897-7_8.

[54] M. Chase, D. Derler, S. Goldfeder, C. Orlandi, S. Ramacher, C. Rechberger, D. Slamanig, and G. Zaverucha, 2017, "Post-Quantum Zero-Knowledge and Signatures from Symmetric-Key Primitives," *Proceedings of the 2017 ACM SIGSAC Conference on Computer and Communications Security*, https://doi.org/10.1145/3133956.3133997.

[55] Y. Ishai, E. Kushilevitz, R. Ostrovsky, and A. Sahai, 2009, Zero-knowledge proofs from secure multiparty computation, *SIAM Journal on Computing* 39(3):1121–1152, https://doi.org/10.1137/080725398.

affect other quantum-resistant signature algorithms. These algorithms may be appropriate for use in specific scenarios such as firmware signing. While their wide application would pose some difficulties for system implementers, they would provide a viable digital signature option for some use cases in the event that a cryptanalytic breakthrough rendered other digital signature algorithms vulnerable.

The NIST effort has strong support from the international cryptographic community and the recommendations that will be published soon are expected to be widely implemented. For implementers, the transition is expected to be difficult and expensive. (See the discussion of the Systems Driver in Chapter 4.) The new candidate post-quantum public-key algorithms currently have longer public keys, longer ciphertexts or signatures, or longer encryption, decryption, or signature verification times compared to public-key algorithms in wide use today. These differences can be significant disadvantages: slower cryptographic operations will require more computational power and slow down network communications; larger keys, ciphertexts, or signatures will require more data transfer and may exceed hard-coded or intrinsic limits on some systems. These differences will result in complicated implementation and transition problems and may result in vulnerabilities in implementations or protocols that hope to use these algorithms as drop-in replacements.

Key distribution centers (KDCs), a technology that can support the distribution of the symmetric key without relying on public-key encryption algorithms, are described in Box 2.3.

FINDING 2.2: For smaller-scale applications within a single sophisticated organization with little tolerance for the possible risks to public-key cryptography (whether from a mathematical advance or quantum computers), it may be possible to use KDCs to replace or augment some uses of public-key cryptography. Because of the different trust models and attack surface, deploying KDCs to replace public-key cryptographic functions in open settings like Hypertext Transfer Protocol Secure (HTTPS) would be difficult technically, politically, and logistically.

Algorithm Transitions

There are several historical examples of widely used cryptographic algorithms being replaced owing to cryptanalytic attacks. These examples can be informative when contemplating the upcoming transition to post-quantum algorithms.

BOX 2.3
Key Distribution Centers

As an alternative to public key distribution systems, key distribution centers (KDCs) use symmetric cryptography to perform authentication and key exchange in systems within a defined setting, serving as a trusted party to perform these actions and providing a method to initialize keys and parties.

The KDC manages a distinct symmetric key shared with each endpoint. When two endpoints want to establish a secure connection, they ask the KDC to issue a new session key, which the KDC encrypts for each endpoint (and no one else).

Before public-key cryptography was widespread, organizations relied on KDC-based encryption schemes at scale. For example, the Department of Defense operated the Bellfield secure voice system. The Kerberos authentication system is still supported in Microsoft Windows and Unix networks.

KDCs are not widely used in open or distributed settings; such an application would face significant engineering and trust problems because of the scale of activity required and would likely have fundamental limitations that increase the system's attack surface, depending on the desired application. Among these are the need for many KDC servers to handle the load, each of which is vulnerable to attack, the difficulty of getting agreement on who should be trusted to operate the KDCs, and the fact that the large scale of the traffic being protected makes the KDCs a very attractive target.

A common challenge is that implementations must often include backward compatibility, even for insecure algorithms, during the transition in order to maintain a functioning system. Users, however, have relatively little impetus to upgrade while backward compatibility works. As a result, insecure algorithms live on even though they are known to be flawed. For example, the MD5 hash function was widely used through the 1990s and early 2000s despite the algorithm containing known flaws; the first collision was demonstrated in 2004, rendering the algorithm insecure, but it has remained in use for years after.[56] A set of academic researchers carried out a proof-of-concept attack using MD5 weaknesses to issue themselves a malicious certificate authority certificate in 2009, and the authors of the Flame malware used an MD5 collision attack against a portion of Microsoft's certificate infrastructure in 2010 in order to create a fake signing certificate for the malware that passed verification. The first better-than-brute-force attack against the SHA-1 hash algorithm was published in 2005, and the first SHA-1 hash collision was demonstrated in 2017, but SHA-1 remains in widespread use.[57]

Another illustrative example is the transition from the severely flawed SSL 2.0. Even though SSL 2.0 was superseded by a new version in 1996, many servers provided backward compatibility support for SSL 2.0. Twenty years later, in 2016, researchers found that hundreds of major websites were vulnerable[58] because they still supported SSL 2.0 and used the same RSA private key with SSL 2.0 and newer versions.[59]

NIST and other organizations are well aware of the challenges posed by algorithm transitions in general, and by the complexity of the anticipated transition to post-quantum encryption algorithms. The Department of Homeland Security has collaborated with NIST on the development of a roadmap for the transition to post-quantum encryption and some private sector organizations have also begun to create and document plans and strategies for managing the transition.[60,61]

The Future Threat to Collected Encrypted Data

When organizations deploy encryption to protect data at rest or in transit, they expect the encryption to protect their data against an adversary that has access to the ciphertext. In this scenario, the encryption needs to be secure against the adversary's current decryption capabilities as well as the adversary's future capabilities for as long as the content of the data is considered sensitive, in the event that the adversary collects the ciphertext and stores it for future analysis.

Providing security guarantees in this scenario means making assumptions about the strength of cryptographic algorithms against future cryptanalytic breakthroughs, evaluations of key length, and predictions for future changes to computer processing power. Breakthroughs in computing such as the emergence of quantum computers or a new approach to classical cryptanalysis have the potential to undermine prior assumptions about the strength of algorithms, and thus the protection afforded to encrypted data captured by an adversary. The risk that encrypted data will be intercepted, saved, and later decrypted has been long understood. In the NSA and Government Communications Headquarters (GCHQ) Venona project, Soviet messages intercepted and archived in the early 1940s continued to be analyzed and broken by the NSA and GCHQ through the late 1970s.[62] Guidance on the use and upgrading of cryptographic systems reflects this understanding of the threat of future decryption and attempts to mitigate that threat.

---

[56] X. Wang and H. Yu, 2005, "How to Break MD5 and Other Hash Functions," pp. 19–35 in *Lecture Notes in Computer Science*, https://doi.org/10.1007/11426639_2.

[57] M. Stevens, E. Bursztein, P. Karpman, A. Albertini, and Y. Markov, 2017, "The First Collision for Full SHA-1," pp. 570–596 in *Advances in Cryptology—CRYPTO 2017*, https://doi.org/10.1007/978-3-319-63688-7_19.

[58] N. Aviram, S. Schinzel, J. Somorovsky, N. Heninger, M. Dankel, J. Steube, L. Valenta, et al., 2016, "DROWN: Breaking TLS Using SSLv2," *25th USENIX Security Symposium*, August.

[59] M. Stevens, E. Bursztein, P. Karpman, A. Albertini, and Y. Markov, 2017, "The First Collision for Full SHA-1," pp. 570–596 in *Advances in Cryptology—CRYPTO 2017*, https://doi.org/10.1007/978-3-319-63688-7_19.

[60] Department of Homeland Security, "Post-Quantum Cryptography," https://www.dhs.gov/quantum, accessed October 5, 2021.

[61] SAFECode, "Post Quantum Crypto Archives," https://safecode.org/category/post-quantum-crypto, accessed October 12, 2021.

[62] NSA, n.d., "The Venona Story," https://www.nsa.gov/portals/75/documents/about/cryptologic-heritage/historical-figures-publications/publications/coldwar/venona_story.pdf.

FINDING 2.3: If an organization has encrypted information in the past using keys negotiated with an algorithm that later becomes vulnerable to cryptanalysis by a quantum or classical computer, there is little that the organization can do at the cryptographic level to prevent future decryption of ciphertexts that have already been intercepted and stored by an adversary. Organizations in that situation may be best served by understanding their risks from decryption of previously encrypted information, assembling an inventory of such information, and taking measures to limit the damage in the event that information is decrypted in the future.

Organizations that are concerned about the security of their data going forward will need to upgrade systems to use quantum-resistant algorithms. National expert agencies such as NIST and NSA in the United States and NCSC in the United Kingdom are tasked with providing relevant algorithm recommendations and standards. These organizations are well aware of the future threat to stored encrypted data through advances in cryptanalysis and take these risks into account when providing recommendations.

EMERGING TECHNOLOGIES

The previous sections described cryptographic algorithms that are widely used today. This section introduces techniques that are not widely adopted, often because they are new, are not well understood, or have technical limitations that might be overcome through future research.

Zero Knowledge Proofs

Zero knowledge (ZK) proofs permit one computer (typically called a "Prover") to convince another computer (typically called a "Verifier") that the Prover knows some fact without revealing the fact to the Verifier.[63,64,65] Zero knowledge is used in cryptocurrency applications to hide the public keys of transaction participants (Zcash).[66,67] More recently, zero knowledge proofs have been used as the basis of post-quantum digital signature algorithms (NIST-Picnic).[68,69]

There have been suggestions that the application of technologies such as zero knowledge can improve citizens' trust in the Intelligence Community and government processes broadly. In general, the committee is skeptical that initiatives by the Intelligence Community based on zero knowledge or other advanced cryptographic methods will create (or substitute for) trust in the Intelligence Community. Fundamentally, trust is a social process, and technical tools cannot replace non-technical foundations of trust. One reason is that mathematical constructions generally do not map precisely to real-world concerns, so cryptographic proofs will not preclude the possibility of Intelligence Community misbehavior. Outside observers will generally lack the expertise, resources, and access required to assess the underlying mathematics, implementation correctness, and operational details to the degree required to establish trust. In contexts where there are conflicting motives (such as privacy regulations impacting

---

[63] S. Goldwasser, S. Micali, and C. Rackoff, 1989, The knowledge complexity of interactive proof systems, *SIAM Journal on Computing* 18(1):186–208, https://doi.org/10.1137/0218012.

[64] M. Blum, P. Feldman, and S. Micali, 1988, "Non-Interactive Zero-Knowledge and Its Applications," *Proceedings of the Twentieth Annual ACM Symposium on Theory of Computing—STOC '88*, https://doi.org/10.1145/62212.62222.

[65] J. Baron, "Securing Information for Encrypted Verification and Evaluation (SIEVE)," DARPA RSS, https://www.darpa.mil/program/securing-information-for-encrypted-verification-and-evaluation, accessed October 22, 2021.

[66] N. Bitansky, A. Chiesa, Y. Ishai, O. Paneth, and R. Ostrovsky, 2013, "Succinct Non-Interactive Arguments via Linear Interactive Proofs," pp. 315–333 in *Theory of Cryptography*, https://doi.org/10.1007/978-3-642-36594-2_18.

[67] E. Ben Sasson, A. Chiesa, C. Garman, M. Green, I. Miers, E. Tromer, and M. Virza, 2014, "Zerocash: Decentralized Anonymous Payments from Bitcoin," *2014 IEEE Symposium on Security and Privacy*, https://doi.org/10.1109/sp.2014.36.

[68] M. Chase, D. Derler, S. Goldfeder, C. Orlandi, S. Ramacher, C. Rechberger, D. Slamanig, and G. Zaverucha, 2017, "Post-Quantum Zero-Knowledge and Signatures from Symmetric-Key Primitives," *Proceedings of the 2017 ACM SIGSAC Conference on Computer and Communications Security*, https://doi.org/10.1145/3133956.3133997.

[69] Y. Ishai, E. Kushilevitz, R. Ostrovsky, and A. Sahai, 2009, Zero-knowledge proofs from secure multiparty computation, *SIAM Journal on Computing* 39(3):1121–1152, https://doi.org/10.1137/080725398.

Intelligence Community mission effectiveness, or where Intelligence Community organizations with offensive roles are making proposals related to defense), skeptics may also suspect intentional backdoors or view proposals as an attempt to divert discussions about topics such as independent oversight or internal culture. On the other hand, these techniques may be useful in the context of interactions between intelligence or defense organizations in the United States or internationally.[70]

Threshold Cryptography

Secret-sharing techniques enable splitting a key or other sensitive data into multiple parts held by different parties or devices, so that qualified sets of participants can reconstruct the secret.[71,72] The secret remains secure if any lesser group misbehaves or is compromised.

Threshold cryptosystems extend this idea to allow two or more participants to cooperate in performing a sensitive computation, where the security and privacy can be preserved even if one or several parties are compromised or act maliciously.[73,74,75,76] For example, threshold systems have been proposed for many other cryptographic primitives, including encryption, signature, and key agreement.[77,78,79]

Computing on Encrypted Data

With most encryption algorithms in use today, the only operation performed on ciphertext is to decrypt it into readable form. For example, when Microsoft's disk encryption feature (BitLocker) protects data at rest or protocols like TLS/SSL protect data in transit, the main cryptographic objective is to ensure that encrypted data cannot be used unless it has been decrypted.

---

[70] See S. Philippe, R. Goldston, A. Glaser, and F. d'Errico, 2016, A physical zero-knowledge object-comparison system for nuclear warhead verification, *Nature Communications* 7:12890, https://doi.org/10.1038/ncomms12890; A. Segal, J. Feigenbaum, and B. Ford, 2016, "Open, Privacy-Preserving Protocols for Lawful Surveillance," July 13, https://arxiv.org/abs/1607.03659; A. Segal, J. Feigenbaum, and B. Ford, "Privacy-Preserving Lawful Contact Chaining, Preliminary Report," http://www.cs.yale.edu/homes/jf/SFF-WPES2016.pdf; A. Bates, K.R.B. Butler, M. Sherr, C. Shields, P. Traynor, and D. Wallach, 2015, Accountable wiretapping-or-I know they can hear you now, *Journal of Computer Security* 23(2):167–195, https://doi.org/10.3233/JCS-140515; and B. Hemenway, S. Lu, R. Ostrovsky, and W. Welser IV, "High-Precision Secure Computation of Satellite Collision Probabilities," in *Security and Cryptography for Networks* (V. Zikas and R. De Prisco, eds.), SCN 2016, Lecture Notes in Computer Science, Volume 9841, Springer, Cham, https://doi.org/10.1007/978-3-319-44618-9_9.

[71] NIST, "Multi-Party Threshold Cryptography," Computer Security Resource Center, Information Technology Laboratory, https://csrc.nist.gov/projects/threshold-cryptography, accessed October 8, 2021.

[72] A. Beimel, 2011, "Secret-Sharing Schemes: A Survey," pp. 11–46 in *Lecture Notes in Computer Science*, https://doi.org/10.1007/978-3-642-20901-7_2.

[73] O. Goldreich, S. Micali, and A. Wigderson, 2019, "How to Play Any Mental Game, or a Completeness Theorem for Protocols with Honest Majority," *Providing Sound Foundations for Cryptography: On the Work of Shafi Goldwasser and Silvio Micali*, https://doi.org/10.1145/3335741.3335755.

[74] M. Ben-Or, S. Goldwasser, and A. Wigderson, 2019, "Completeness Theorems for Non-Cryptographic Fault-Tolerant Distributed Computation," *Providing Sound Foundations for Cryptography: On the Work of Shafi Goldwasser and Silvio Micali*, https://doi.org/10.1145/3335741.3335756.

[75] D. Chaum, C. Crépeau, and I. Damgård, 1988, "Multiparty Unconditionally Secure Protocols," *Proceedings of the Twentieth Annual ACM Symposium on Theory of Computing—STOC '88*, https://doi.org/10.1145/62212.62214.

[76] R. Cramer, I.B. Damgård, and J.B. Nielsen, 2015, *Secure Multiparty Computation and Secret Sharing*, Cambridge, UK: Cambridge University Press.

[77] D. Boneh, X. Boyen, and S. Halevi, 2006, "Chosen Ciphertext Secure Public Key Threshold Encryption Without Random Oracles," pp. 226–243 in *Topics in Cryptology—CT-RSA 2006*, https://doi.org/10.1007/11605805_15.

[78] V. Shoup, 2000, "Practical Threshold Signatures," pp. 207–220 in *Advances in Cryptology—EUROCRYPT 2000*, https://doi.org/10.1007/3-540-45539-6_15.

[79] D. Harnik, J. Kilian, M. Naor, O. Reingold, and A. Rosen, 2005, "On Robust Combiners for Oblivious Transfer and Other Primitives," pp. 96–113 in *Lecture Notes in Computer Science*, https://doi.org/10.1007/11426639_6.

Some new cryptographic techniques, including fully homomorphic encryption (FHE) and secure multi-party computation (MPC), are different.[80,81,82,83] FHE allows a participant to perform arbitrary computation on encrypted data. MPC allows two or more participants to compute certain information on joint data without ever disclosing this data either to outside adversaries or even to each other. Such techniques could allow organizations to gain intelligence across siloed datastores that cannot be joined.[84,85,86,87,88,89] In theory, entirely untrusted cloud computers could perform general-purpose calculations without ever being able to see the underlying data.

FHE allows a computer to perform computations on encrypted data generating encrypted answers; the answers are available only to an entity that has the decryption key. MPC allows groups of machines to jointly compute functions of secret-shared data. The result of an MPC is the secret-shared output of the computation, which can then be sent to one or several parties for reconstruction. Many MPC protocols also have the benefit that even if a subset of parties misbehave, the computation is still performed correctly.[90]

With present constructions, MPC involves more communication than FHE but is several orders of magnitude more efficient. While FHE is currently about 10,000 to 100,000 times slower than unencrypted computation[91] and works only in the so-called circuit model of computation (where programs must be converted to digital circuits with addition and multiplication "gates"[92]), MPC has progressed much further in performance, and is often a far better alternative to FHE. For example, as of 2012, MPC could support a million operations per second on encrypted data.[93] Since 2012, MPC research has progressed even further and performance at the level of 10×–100× slowdown compared to unencrypted computation can be achieved in some cases. For MPC, there are benchmarks for database operations that are 5 to 10 times slower than SQL.[94] Unlike FHE, there are MPC techniques to support random access into an array in constant, rather than linear time.[95] The state of the art in zero knowledge proofs

---

[80] See the Palisade Homomorphic Encryption Software Library website at https://palisade-crypto.org, accessed October 8, 2021.

[81] S. Halevi and V. Shoup, 2020, "Design and Implementation of HElib: A Homomorphic Encryption Library," Cryptology ePrint Archive, https://eprint.iacr.org/2020/1481.

[82] EMP, "Toolkit: EMP-Tool," https://emp-toolkit.github.io/emp-doc/html/md___users_wangxiao_git_emp-toolkit_emp-readme__r_e_a_d_m_e.html, accessed October 8, 2021.

[83] KU Leuven, "SCALE-MAMBA Software," https://homes.esat.kuleuven.be/~nsmart/SCALE, accessed October 11, 2021.

[84] Boston Women's Workforce Council, "Data Privacy," https://thebwwc.org/mpc, accessed October 8, 2021.

[85] HAIC Helsinki-Aalto Institute for Cybersecurity, 2018, "HAIC Talk: The Advertisement Exchange—with Moti Yung," YouTube, August 2, https://www.youtube.com/watch?v=bMl2PFU7gMA.

[86] A. Walker, S. Patel, and M. Yung, 2019, "Helping Organizations Do More Without Collecting More Data," Google Security Blog, June 19, 2019, https://security.googleblog.com/2019/06/helping-organizations-do-more-without-collecting-more-data.html.

[87] Apple, 2021, "Apple and Google Partner on COVID-19 Contact Tracing Technology," Newsroom, August 6, https://www.apple.com/newsroom/2020/04/apple-and-google-partner-on-covid-19-contact-tracing-technology.

[88] M. Ion, B. Kreuter, A.E. Nergiz, S. Patel, S. Saxena, K. Seth, M. Raykova, D. Shanahan, and M. Yung, 2020, "On Deploying Secure Computing: Private Intersection-Sum-with-Cardinality," pp. 370–389 in *2020 IEEE European Symposium on Security and Privacy (EuroS&P)*, https://doi.org/10.1109/eurosp48549.2020.00031.

[89] F.S. Dittmer, Y. Ishai, S. Lu, R. Ostrovsky, M. Elsabagh, N. Kiourtis, B. Schulte, and A. Stavrou, 2020, "Function Secret Sharing for PSI-CA: With Applications to Private Contact Tracing," CoRR abs/2012.13053.

[90] V. Goyal, H. Li, R. Ostrovsky, A. Polychroniadou, and Y. Song, 2021, "Atlas: Efficient and Scalable MPC in the Honest Majority Setting," pp. 244–274 in *Advances in Cryptology—CRYPTO 2021*, https://doi.org/10.1007/978-3-030-84245-1_9.

[91] A. Feldmann, N. Samardzic, A. Krastev, S. Devadas, R. Dreslinski, K. Eldefrawy, N. Genise, C. Peikert, and D. Sanchez, 2021, "F1: A Fast and Programmable Accelerator for Fully Homomorphic Encryption (Extended Version)," revised September 25, https://arxiv.org/abs/2109.05371.

[92] A gate is a logical operation, such as addition or multiplication of two numbers, or a Boolean operation such as a logical and or a logical or. It is well known in computer science that any computer program can be represented as a collection of elementary gates. Computer hardware implemented in silicon consists of logical gates that can do elementary computation, and computers put these operations together to run arbitrary computer programs. Unfortunately, in cryptography, one must do the "reverse" and decompose computer programs into logical gates, as computing on encrypted data in FHE requires. For MPC, there is more flexibility about the model of computation.

[93] D. Bogdanov, M. Niitsoo, T. Toft, and J. Willemson, 2012, High-performance secure multi-party computation for data mining applications, *International Journal of Information Security* 11:403–418, https://link.springer.com/article/10.1007/s10207-012-0177-2.

[94] I. Yuval, E. Kushilevitz, S. Lu, and R. Ostrovsky, 2016, "Private large-scale databases with distributed searchable symmetric encryption." In Cryptographers' Track at the RSA Conference, pp. 90–107. Springer, Cham.

[95] R. Ostrovsky and V. Shoup, 1997, "Private information storage (extended abstract)." *Symposium on Theory of Computing (STOC)* '97.

is even more encouraging, and implementation can achieve 3×–5× slowdown compared to unencrypted execution of the same program.[96]

There are special cases of MPC that, instead of supporting general computation, support only a specific computing task but are even more efficient than general-purpose MPC. One example is Function Secret Sharing (FSS), which allows two (or more) servers to compute simple aggregate statistics with privacy of input and output from the computing servers.[97,98,99,100] FSS is being deployed (as part of Prio) in the Mozilla Firefox Browser to compute users' telemetry data in a privacy-preserving way. Two other examples of special-purpose MPC are so-called private set intersection (PSI) and private set union (PSU). Private set intersection allows two or more parties to find intersections of their secret sets of data without learning anything else, while private set union allows parties to find a union without finding out who contributed a specific item.

PSI has many applications in practice—for example, to manage allow lists or block lists privately, to privately check whether a password has been leaked, and to privately match ads displayed on the web to users. There have been proposals to use PSI to carry out law enforcement searches for user information without revealing non-matching data.[101,102,103] Google currently uses PSI to calculate statistics on the overlapping membership of data sets without sharing the underlying data sets.

There are also faster but limited-functionality variants of homomorphic encryption that are referred to as partially (or somewhat) homomorphic encryption. These variants, which for example support performing computations of limited complexity or using only addition operations, can be sufficient for tasks such as cryptographic tallying of votes and computation of statistics.[104] To give an idea of the current state of the art for a well-suited application, a 2020 proof-of-concept somewhat homomorphic computation of a "genome-wide association study" that involved computing a logistic regression homomorphically extrapolated a running time of 23 hours and several terabytes of RAM for a sample of 10,000 genomes and 500,000 markers;[105] a 2013 publication[106,107,108] suggested that an equivalent computation was doable in 7 minutes on a laptop.

FINDING 2.4: The research community continues to make improvements in the technology of computation on encrypted data. Such improvements can be expected to enable new ways of securely sharing both government and private-sector information.

---

[96] S. Dittmer, Y. Ishai, and R. Ostrovsky, 2020, "Line-point zero knowledge and its applications." Cryptology ePrint Archive.

[97] E. Boyle, N. Gilboa, and Y. Ishai, 2016, "Function Secret Sharing." *Proceedings of the 2016 ACM SIGSAC Conference on Computer and Communications Security*, https://doi.org/10.1145/2976749.2978429.

[98] H. Corrigan-Gibbs and D. Boneh, 2017, "Prio: Private, Robust, and Scalable Computation of Aggregate Statistics," pp. 259–282 in *14th USENIX Symposium on Networked Systems Design and Implementation*, https://www.usenix.org/system/files/conference/nsdi17/nsdi17-corrigan-gibbs.pdf.

[99] S. Englehardt, 2019, "Next Steps in Privacy-Preserving Telemetry with Prio," Mozilla Security Blog, June 6, https://blog.mozilla.org/security/2019/06/06/next-steps-in-privacy-preserving-telemetry-with-prio.

[100] S. Addanki, K. Garbe, E. Jaffe, R. Ostrovsky, and A. Polychroniadou, 2021, "Prio+: Privacy Preserving Aggregate Statistics via Boolean Shares," Cryptology ePrint Archive, https://eprint.iacr.org/2021/576.

[101] J. Feigenbaum and B. Ford, 2015, Seeking Anonymity in an Internet Panopticon, *Communications of the ACM* 58(10):58–69, https://doi.org/10.1145/2714561.

[102] P. Rindal and P. Schoppmann, 2021, "Vole-Psi: Fast OPRF and Circuit-Psi from Vector-OLE," pp. 901–930 in *Annual International Conference on the Theory and Applications of Cryptographic Techniques*, Springer.

[103] M. Rosulek and N. Trieu, 2021, "Compact and Malicious Private Set Intersection for Small Sets," pp. 1166–1181 in *Proceedings of the 2021 ACM SIGSAC Conference on Computer and Communications Security*.

[104] P. Paillier, 2002, Composite-residuosity based cryptography: An overview, *CryptoBytes* 5(1).

[105] M. Blatt, A. Gusev, Y. Polyakov, and S. Goldwasser, 2020, Secure large-scale genome-wide association studies using homomorphic encryption, *Proceedings of the National Academy of Sciences U.S.A.* 117(21):11608–11613, https://www.pnas.org/content/117/21/11608.

[106] K. Sikorska, E. Lesaffre, P.F.J. Groenen, and P.H.C. Eilers, 2013, GWAS on your notebook: Fast semi-parallel linear and logistic regression for genome-wide association studies, *BMC Bioinformatics* 14:166, https://bmcbioinformatics.biomedcentral.com/articles/10.1186/1471-2105-14-166.

[107] P. Rindal and P. Schoppmann, 2021, "Vole-Psi: Fast OPRF and Circuit-Psi from Vector-OLE," pp. 901–930 in *Annual International Conference on the Theory and Applications of Cryptographic Techniques*, Springer.

[108] M. Rosulek and N. Trieu, 2021, "Compact and Malicious Private Set Intersection for Small Sets," pp. 1166–1181 in *Proceedings of the 2021 ACM SIGSAC Conference on Computer and Communications Security*.

Encrypted Data Storage and Retrieval

Specialized cryptographic techniques have been developed to help address security challenges involved with storing, searching, and retrieving information.[109,110,111] Searchable encryption refers to techniques that allow participants to perform limited kinds of queries on an encrypted database while concealing the contents of the database and the search queries from the computers hosting it. Oblivious Random Access Memory (ORAM) is a technique that can hide from a storage system which indices or files have been accessed or accessed more than once.[112] ORAM is used, for example, in the Signal messaging platform to perform contact discovery without revealing a user's contact list to the untrusted machine performing the computation.[113]

Private Information Retrieval (PIR) is a technology that allows a user to send an encrypted query to a database, which "processes" the entire database and gives back a short, encrypted answer.[114,115,116] The database system that executes the query cannot determine what the query is. Although initial PIR schemes have been very inefficient, more efficient approaches involving off-line precomputation may be more practical.[117,118] These techniques allow one to select data to be uploaded to a more sensitive processing environment without revealing which data are being uploaded.

Program Obfuscation

In theory, a cryptographically secure method for program obfuscation would allow an entity to distribute a software program for performing some function (e.g., decrypting a ciphertext using a fixed secret key) that could be executed by a third party without revealing any information about the program's execution (e.g., the value of the key used to decrypt) beyond the result.[119]

Engineering Alternatives to Cryptography's Limitations

For many real-world data security problems, there are no purely cryptographic solutions, or the available cryptographic algorithms do not meet real-world efficiency constraints. For example, the existing cryptographically robust proposals for program obfuscation are so slow as to be unimplementable for real-world applications, but there is a market need for practical solutions in digital rights management and other applications. The

---

[109] D. Boneh, G. Di Crescenzo, R. Ostrovsky, and G. Persiano, 2004, "Public Key Encryption with Keyword Search," pp. 506–522 in *Advances in Cryptology—EUROCRYPT 2004*, https://doi.org/10.1007/978-3-540-24676-3_30.

[110] R. Curtmola, J. Garay, S. Kamara, and R. Ostrovsky, 2011, Searchable symmetric encryption: Improved definitions and efficient constructions, *Journal of Computer Security* 19(5):895–934, https://doi.org/10.3233/jcs-2011-0426.

[111] Intelligence Advanced Research Projects Agency, 2010, "SPAR: Security and Privacy Assurance Research," Broad Agency Announcement IARPA-BAA-11-01, released December 29, https://www.iarpa.gov/index.php/research-programs/spar.

[112] O. Goldreich and R. Ostrovsky, 1996, Software protection and simulation on oblivious rams, *Journal of the ACM* 43(3):431–473, https://doi.org/10.1145/233551.233553.

[113] Moxie0, 2017, "Technology Preview: Private Contact Discovery for Signal," *Signal* (blog), September 26, https://signal.org/blog/private-contact-discovery.

[114] B. Chor, E. Kushilevitz, O. Goldreich, and M. Sudan, 1998, Private information retrieval, *Journal of the ACM* 45(6):965–981, https://doi.org/10.1145/293347.293350.

[115] E. Kushilevitz and R. Ostrovsky, 1997, "Replication Is Not Needed: Single Database, Computationally-Private Information Retrieval," *Proceedings 38th Annual Symposium on Foundations of Computer Science*, https://doi.org/10.1109/sfcs.1997.646125.

[116] R. Ostrovsky and W.E. Skeith, 2007, "A Survey of Single-Database Private Information Retrieval: Techniques and Applications," pp. 393–411 in *Public Key Cryptography—PKC 2007*, https://doi.org/10.1007/978-3-540-71677-8_26.

[117] H. Corrigan-Gibbs and D. Kogan, 2020, "Private Information Retrieval with Sublinear Online Time," pp. 44–75 in *Advances in Cryptology—EUROCRYPT 2020*, https://doi.org/10.1007/978-3-030-45721-1_3.

[118] E. Shi, W. Aqeel, B. Chandrasekaran, and B. Maggs, 2021, "Puncturable Pseudorandom Sets and Private Information Retrieval with Near-Optimal Online Bandwidth and Time," pp. 641–669 in *Advances in Cryptology—CRYPTO 2021*, https://doi.org/10.1007/978-3-030-84259-8_22.

[119] S. Garg, C. Gentry, S. Halevi, M. Raykova, A. Sahai, and B. Waters, 2016, Candidate indistinguishability obfuscation and functional encryption for all circuits, *SIAM Journal on Computing* 45(3):882–929, https://doi.org/10.1137/14095772x.

implementations that have been deployed—called "white box" cryptography—lack mathematical foundations for security, and typically get broken rapidly.

Another market need is for secure computation at speeds competitive with conventional microprocessors. The products that have been built to meet this need include specialized modes such as Intel's software guard extensions (SGX) in general-purpose processors, as well as specialized tamper-resistant microprocessors such as those found in smart cards and mobile phones. As discussed in the Systems section of Chapter 4, these products can be broken if there are imperfections in the tamper resistance or bugs in the underlying implementations.

Quantum Key Distribution

Quantum key distribution (QKD) aims to establish secret keys for communications using quantum mechanical properties of photons, rather than the mathematical problems used in public-key cryptography. The practical applicability of the approach is limited to scenarios where senders and receivers directly exchange photons, such as over a private fiber connection.[120] As a result, QKD cannot be used on typical packet switched networks (such as the Internet), radio connections (such as Wi-Fi or mobile phones), or copper wire (Ethernet). As QKD does not provide any form of authentication, entities must authenticate each other by classical techniques such as prenegotiated secret keys or public-key technology. In addition, QKD trades a relatively well studied problem (implementing the mathematical operations involved in public-key cryptography) against less understood engineering problems associated with building secure optical equipment. Nevertheless, commercial systems have been offered for sale, and China operates an experimental satellite-based QKD system that has received much attention, including U.S. government plans for a competing system.

QKD may have niche applications in scenarios that require an extra layer of security against cryptanalytic attacks, and that can live within the limitations of its highly constrained quantum photonic channel. However, the fundamental limitations of QKD mean that, even with research or engineering advances, the approach will not play a significant role in the overall data security landscape.[121]

---

[120] See a recent survey, P. Sharma, A. Agrawal, V. Bhatia, S. Prakash, and A.K. Mishra, 2021, Quantum key distribution secured optical networks: A survey, *IEEE Open Journal of the Communications Society* 2:2049–2083, http://doi.org/10.1109/OJCOMS.2021.3106659.

[121] NSA, 2021, "Quantum Computing and Post-Quantum Cryptography," PP-21-1120, August, https://media.defense.gov/2021/Aug/04/2002821837/-1/-1/1/Quantum_FAQs_20210804.PDF.

3

Methodology

The statement of task directs the committee to "identify potential scenarios for the balance during the next 10 to 20 years between encryption and decryption (and other data and communications protection and exploitation capabilities) and assess the national security and intelligence implications of each scenario." In particular, the committee is required to:

- Identify the drivers of the future, design and select scenarios for exploration, and consider the implications for the application of encryption and decryption capabilities, as well as the possible pace and effects of the adoption of post-quantum cryptology.
- Assess the national security, intelligence, and broad societal implications of each scenario determined by the committee to be most worthy of attention; identify and assess options for responding to these scenarios; and assess the implications for future Intelligence Community science and technology and security investments.

For the purposes of this study, the committee used a tailored mix of approaches based on the work of futures and strategic foresight firms, such as Toffler Associates and academic researchers such as Steven Weber at the Center for Long-Term Cybersecurity.[1,2] The committee followed a five-step process:

1. Develop the focal question.
2. Scan the environment.
3. Identify the drivers of change.
4. Design and select the scenarios.
5. Identify and explore strategically salient issues.

---

[1] G.S. Parnell, J.A. Jackson, R.C. Burk, L.J. Lehmkuhl, and J.A. Engelbrecht, Jr., 1999, R&D concept decision analysis: Using alternate futures for sensitivity analysis, *Journal of Multi-Criteria Decision Analysis* 8(3):119–127.

[2] J.A. Englebrecht, R.L. Bivens, P.M. Condray, M.D. Fecteau, J.P. Geis II, and K.C. Smith, 1996, *Alternate Futures for 2025*, Air University Press, Maxwell Air Force Base, AL.

DEVELOP THE FOCAL QUESTION

The focal question establishes foundation and context for the study and provides key insights into the required analysis. It is important to note that, in following the methodologies selected by the committee, the focal question that guides the effort may not necessarily be the same as what is found in the statement of task. In this case, while the sponsor outlined its objectives and supporting tasks, there was no single overarching question that would scope and guide the committee's efforts. For any investigation of future scenarios, a focal question is critical to ensure both an efficient use of resources and a valuable outcome. Every experiment must start with a question, and this effort is no different; getting the question "right" is a critical first step in ensuring a valuable outcome for any futures and foresight effort.

The committee began by reviewing the questions and considerations identified by the sponsor in the statement of task (see Chapter 1) and determining if there was a way to aggregate them into a single, overarching question. After internal deliberation and research, the committee settled on the following focal question to guide the effort: *What opportunities and risks associated with encryption and decryption will emerge over the next 10 to 20 years, and how can the Intelligence Community prepare itself?*

As the committee began the task, the sponsor provided additional questions to help focus the work. These are presented in Box 3.1.

BOX 3.1
Questions and Considerations from the Sponsor
That Informed the Focal Question

- Potential alternative technological developments during the next decade that would affect the balance between "encryption" and "decryption and non-decryption data exploitation schemes" in the 10- to 20-year time frame. Examples of alternative developments might include assuming that the post-quantum cryptography (PQC) becomes widely available before Cryptographically Relevant Quantum Computers (CRQCs); assuming that CRQCs or other means of exploiting protected data become widely available before appropriately resistant cryptology; or assuming that PQC and other resistant cryptology become widespread only shortly before (within 5 years) of CRQCs or other means of exploiting protected data.
- Most likely sources for surprising progress and burst of innovation in "encryption/data protection or decryption/data exploitation capabilities" in the next 10 years.
- The reactions of other governments in each of these scenarios, especially if advanced encryption and decryption techniques are available only to a small number of countries.
- Which scenarios appear most likely and least likely based on current trends?
- The signposts for each scenario that would reduce uncertainty about the scenario coming true.
- The "game changers" that would alter current expectations for the future balance between encryption and decryption capabilities and what would be the indicators that such a shift was approaching.
- Recent research and technology developments, including foreign and commercial, that have occurred after the National Academies' study on quantum computing.
- Development and timelines for new encryption capabilities, such as user-controlled encryption, lattice-based cryptology, code-based cryptology, supersingular isogenies, and hash-based signatures, and capabilities for computing on encrypted data.
- The implications of further developments in non-decryption-based schemes, including side-channel attacks and metadata analysis.
- The implications of the implementation of new-government regulations to gain access to commercial encrypted data.
- The potential application of artificial intelligence and machine learning to cryptology problems.
- The implications of the proliferation of the Internet of Things (IoT), lightweight encryption, embedded devices, and "safe city" technologies to the future encryption-decryption environment.

SCAN THE ENVIRONMENT

Although scenario development involves some imagination, scenarios must be grounded in data. Once the committee identified the focal question, it was able to identify primary and secondary sources and receive presentations from leading experts across relevant fields related to cryptography to ensure a broad understanding of the issues at play. This environmental scan allowed the committee to start identifying the *features* that would influence the future operating environment in its scenarios. The identification of a feature aims to answer the question, What might be out there?—features are attributes or characteristics of the future operating environment, or the material objects expected to be present in it, based on the information available (because, when looking to 2030 and 2040, it is difficult to find "reliable" forecast data). Features are considered data, but at their foundation they are the effects of human invention and behavior from the environment.

Having identified the potential features of the future operating environment, the committee could begin to explore their implications—that is, to ask the question, "Given the features, what is likely to emerge in the future?" More than cursory trends, the exploration of how features will evolve and emerge in the explored future allowed the committee to identify extrapolated, and linear (or faster) change in various categories subject to quantification or qualification when considered by themselves. The implications of the features are effects or results, and they need not be physical or tangible, allowing a more nuanced approach to imagining potential future scenarios.

Although the question focused on encryption, it was important for the committee to consider the global societal and political context in which encryption would be applied. This necessitated exploring such features as culture, economics, industry, intelligence, law, politics, regulatory environment, and society.

IDENTIFY THE DRIVERS OF CHANGE

What Are "Drivers"?

Following the Toffler Associates' process,[3] the committee assigned its observations to three *drivers* of the future operating environment. *Drivers* are defined as material or immaterial forces or vectors expected to be a significant cause of or contributor to change. Each of the three drivers is multi-dimensional and wide-ranging in its application within the scope of the focal question. Box 3.2 highlights important points using drivers.

Each of the three drivers is framed by two endpoints—they are the extreme values at each end of the driver, and they bound the range of possibility of each driver. With each of the three drivers defining an orthogonal axis, one can envision the three-dimensional "decision space" defined by the drivers and bounded by the endpoints. The extremes of the three drivers define eight corners of a cube, creating eight unique potential scenarios for exploration.

Creating and Naming Drivers and Endpoints

To develop the drivers, the committee reviewed the data from the environmental scan, the members' understanding of encryption and related policy and technology issues, and the issues raised by the statement of task, and discussed the features and implications that emerged. Through a series of facilitated discussions, the committee

---

BOX 3.2
Drivers Developing Scenarios

Each driver has two endpoints and, where those endpoints converge, creates a unique scenario—three drivers create eight scenarios.

---

[3] The Toffler Associates' process identifies three drivers that encompass the decision space for the future operating environment. Three drivers is a process/methodology goal that defines a cube representing the decision space of outcomes along each axis.

arrayed the information it had collected from the previous environmental scan, introduced additional relevant information, and identified three groupings. While exploring the logic for each grouping, the committee worked to identify the major forces that the groupings represented—these forces would then serve as natural names for the drivers. Naming drivers can be difficult, because one word or phrase may not capture the breadth of features and implications, but the committee sought to select names for the drivers and endpoints that would enable the reader to understand their intent quickly. In the process of developing and exploring the drivers, the committee also identified a set of findings and actions that would be broadly beneficial that appeared to be consequences of the drivers themselves rather than their interactions in scenarios. The latter are not being presented as formal recommendations, however.

For the purposes of this report, the committee identified and named the three drivers as follows: Scientific Advances, Society and Governance, and Systems. Figure 3.1 provides a short description of the drivers and how they bound the decision space.

DESIGN AND SELECT SCENARIOS

To design the scenarios, the committee first sketched a narrative for each of the eight scenarios, leveraging data from the environmental scan and making inferences from the interactions of the drivers. Like the creation of the drivers and endpoints, the narrative for each of the eight scenarios involves a mixture of imagination and analytical rigor—although they might not necessarily be *probable*, they must be both *possible* and *plausible*. Possible because no scenario should necessarily break the accepted laws of physics; plausible because they should allow decision makers to see themselves (or a future version of themselves) in the scenario to maximize immersion and learning.

Selecting the Scenarios

Based on the convergence of the endpoints for the drivers (see Appendix C), the committee discussed what each scenario might look and feel like in the world of 2040 and how each might be relevant to the discussion. After the committee created a thumbnail sketch of each of the eight scenarios, it set out to select the scenarios that it believed would be most interesting to explore the focal question and drive results for the sponsor.

The objective of the committee was to select three scenarios that would be stressful for planning and pose a wide range of situations, as well as meet criteria derived from the focal question. Other criteria the committee used in selecting the scenario:

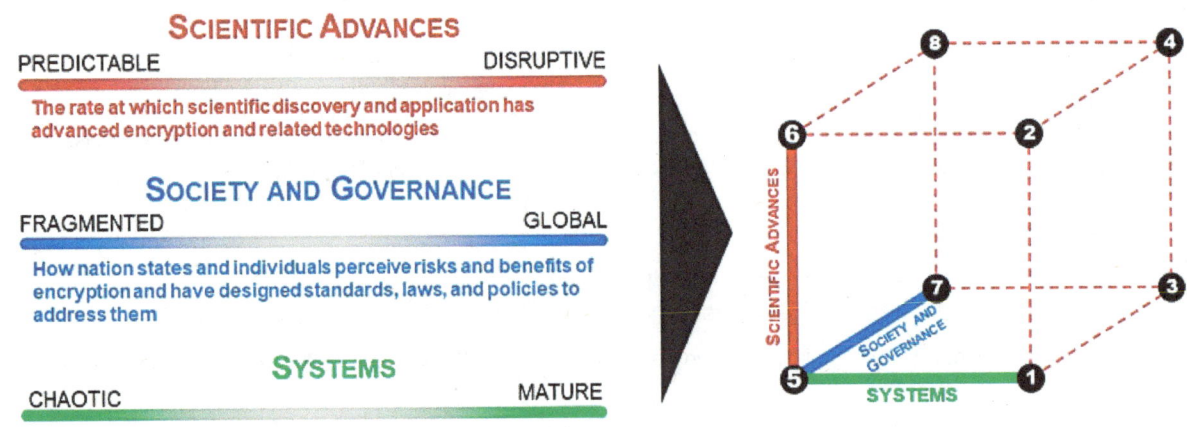

FIGURE 3.1 Visual representation of space defined by drivers.

- One future should be *similar* to the world in which we find ourselves today—that is, the result of extrapolating established trends. This scenario is not intended to be today—rather, it is a future state when many things have changed, but this scenario will be perceived as *like* today.
- One future should be considered the most stressful and challenging world to explore, given the focal question. This future may be the one farthest away in the cube from the status quo.
- One future should fall between those two extremes, allowing for the exploration of different alternatives.

Developing the Narratives

Based on its discussions and the criteria, the committee selected three scenarios that are defined and discussed in Chapter 5. Each of the scenarios selected by the committee presents varying degrees of complexity to explore the future of encryption and the impacts on the Intelligence Community. Once the committee selected the three scenarios that it would explore, it set out to design detailed narratives that would help to capture the relevant aspects of each. The descriptions sought to capture the most salient aspects of the future and answer questions such as, What nations are allies and adversaries? How do nations interact? How do people use the Internet? What is the role of the government in daily life? What about crime levels, education, technological advances, economy, and the regulatory environment?" The purpose of each narrative is to provide enough detail to provide a reader with a clear picture of the scenario, to tell a compelling story, and to help explore the questions, What is the sponsor's biggest challenge or concern in this scenario? What is its biggest opportunity?

IDENTIFY AND EXPLORE STRATEGICALLY SALIENT ISSUES

The selected scenarios should identify and elevate issues for consideration. Some issues seem like imperatives and will need to be addressed because they are vital to the sponsor's success in all or most of the scenarios; some will be important, but relevant in only one or two scenarios. Conversely, some future scenarios suggest alternative eventualities whose occurrence (or failure to occur) should be assessed at some point in the future.

The committee documented risks, opportunities, and actions associated with each scenario. In addition, the committee assembled consolidated findings and observations associated with the drivers that, in the committee's view, applied no matter which scenario or individual developments occur in the future.

4

Drivers

The interactions among society, technology, and cryptography have continually evolved, driven by such trends as the move to cloud computing, networked utilities, and the Internet of Things (IoT). There is no reason to believe that such evolution will cease in coming decades. This section describes the three drivers whose state the committee assessed would largely determine the future of encryption and as appropriate explores their interaction with current and foreseeable future technological and societal trends. The description of each driver begins with a summary chart that identifies the endpoints for that driver and lists some of the most significant developments in or attributes of a future that is driven by that extreme. For each driver, the summary chart is followed by an in-depth discussion of the driver and the characteristics and implications of its extremes.

As discussed in Chapter 3, each of the drivers deals with a different aspect of the future being explored. The committee determined that it was most appropriate to describe each driver in a format that seemed best aligned with the components and variables that make up that driver. Thus, although the summary charts for the three drivers have a common style and format, the detailed descriptions of the drivers and their endpoints vary depending on the nature and components of that driver.

SCIENTIFIC ADVANCES DRIVERS: DISRUPTIVE VERSUS PREDICTABLE

Scientific Advances deals with the emergence of new theoretical breakthroughs or significant technologies that impact cryptography. The creation of a large-scale fault-tolerant quantum computer would be one such advance, but new mathematical attacks on asymmetric encryption, advances that enable efficient computation on encrypted data, and technologies that use quantum properties for encryption also fall into this category. At two extremes, the state of future scientific advances could be either *predictable* or *disruptive* (see Figure 4.1).

The Scientific Advances driver measures the rate at which scientific discovery and application advance encryption and related technologies.

Scientific advances are a necessary precursor for technology to have impact in practice, but they are not sufficient on their own. As in all scientific fields, only a relatively small fraction of research results in cryptography are ultimately practical. It can often take years and additional discoveries to figure out what is useful and for ideas to be developed enough to have impact in the real world. At a very high level, a path to practicality for a scientific advance might begin with work giving the theoretical foundations for an idea, followed by work that fleshes out the idea more concretely, a series of academic-scale, proof-of-concept implementations to demonstrate real-world

Driver Details: Scientific Advances

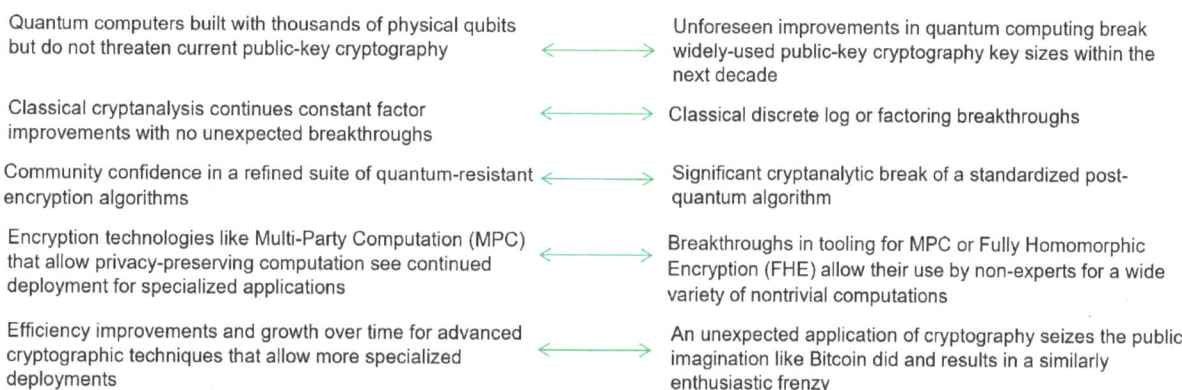

FIGURE 4.1 Visual representation of space defined by Scientific Advances driver.

practicality, limited use in special-purpose applications by industrial research, standardization, incorporation into widely used devices or libraries, and use in general-purpose applications. Scientific advances can also follow practice, as when real-world discoveries of new classes of vulnerabilities (weaknesses) spur research in the underlying causes, or the development and deployment of new technologies (like graphics processing units [GPUs], secure enclaves, or cryptocurrencies) can motivate researchers to develop new theories about the security strengths and weaknesses of technologies and systems.

The other two drivers attempt to capture the conditions that might promote or discourage the transfer of research progress into practice. Research and development often proceed at an uneven pace: ideas can be introduced and languish unnoticed for years, fads and fashionable areas can produce large quantities of publications in niche areas all at once, and there can be communication gaps between the research community and practitioners.

The cryptography research community is broad and spans a range from purely theoretical computer science and mathematics to applied security and privacy work with an implementation focus. As in many academic fields, the vast majority of published papers are theoretical, and it is often hard to predict if a given work will have an impact in practice. There are also research areas that, despite large amounts of research investment and large numbers of academic publications, may never become practical enough to have impact on the real world (e.g., theoretical work on provable software obfuscation), and appear destined to remain of theoretical or only academic interest for the foreseeable future.

The endpoints of the Scientific Advances driver are "predictable" at one extreme, and "disruptive" at the other. "Predictable" scientific advances means that the committee envisions progress to continue at the current pace, with no great surprises to researchers working in the field. "Disruptive" advances would include some kind of major scientific development that is both unexpected given the pace and direction of current results and likely to have a significant impact on either theory or practice.

The committee includes both offensive and defensive advances under this umbrella. Cryptography is somewhat unusual among mathematical and scientific fields in that a "scientific advance" can be offensive or destructive and show that a construction or an idea is in fact insecure, even if it was previously believed to be secure. This is because the security of a cryptographic scheme usually relies on the assumption that a particular computational problem is hard for an adversary to solve in a feasible amount of time. For problems useful for cryptography, it is rare and typically applies only in limited circumstances to be able to specify fully rigorous lower bounds for the computational hardness of problems. Confidence in the hardness of problems is typically gained after years of public analysis has failed to produce improved algorithms for solving them.

The following sections describe the endpoints of "predictable" and "disruptive" advances by laying them out in several relevant areas.

Quantum Computing and Cryptanalysis

The 2019 National Academies' report *Quantum Computing: Progress and Prospects* is a detailed study of the past, present, and future of quantum computing.[1] The authors did not give explicit predictions for a timeline for the development of quantum computing technology. They stated in their summary:

> Predicting the future is always risky, but it can be attempted when the product of interest is an extrapolation of current devices that does not span too many orders of magnitude. However, to create a quantum computer that can run Shor's algorithm to find the private key in a 1024-bit RSA encrypted message requires building a machine that is more than five orders of magnitude larger and has error rates that are about two orders of magnitude better than current machines, as well as developing the software development environment to support this machine.
>
> The progress required to bridge this gap makes it impossible to project the time frame for a large error-corrected quantum computer, and while significant progress in these areas continues, there is no guarantee that all of these challenges will be overcome. The process of bridging this gap might expose unanticipated challenges, require techniques that are not yet invented, or shift owing to new results of foundational scientific research that change our understanding of the quantum world. Rather than speculating on a specific time frame, the committee identified factors that will affect the rate of technology innovation and proposed two metrics and several milestones for monitoring progress in the field moving forward.

The conclusions of that report are still valid as the committee writes this report 2 years later.

Predictable

According to the Global Risk Institute's Quantum Threat Timeline Report 2020, around half of the polled experts think there is a 5 percent or less chance of a "significant quantum threat to public-key" cryptography within the next 10 years, but 85 percent of the experts believed there was a 50 percent or greater chance of a threat to public-key cryptography within 20 years. A useful quantum computer would implement logical "qubits" (see Chapter 2) as the basis of its computational power. In the cited survey, most respondents believed that construction of a single logical qubit (a qubit with reliable error correction) would be demonstrated within 1 to 3 years. In order to threaten current public-key cryptography, a quantum computer with thousands of logical qubits would need to be constructed. Current quantum computers have been built with under a hundred physical qubits ("noisy" qubits without error correction), and an extrapolation of this progress would have research prototypes built with thousands of physical qubits by the end of the decade. At current estimates, a quantum computer would need 10 million to 20 million physical qubits to factor 2048-bit RSA.[2,3] Consistent progress in scaling quantum computers may encounter yet unknown engineering challenges, and expert opinion currently differs on precisely when such a quantum computer would be built. The authors of the *Quantum Computing* report identified the constant average

---

[1] National Academies of Sciences, Engineering, and Medicine (NASEM), 2019, *Quantum Computing: Progress and Prospects*, Washington, DC: The National Academies Press, https://doi.org/10.17226/25196.

[2] V. Gheorghiu and M. Mosca, 2019, "Benchmarking the Quantum Cryptanalysis of Symmetric, Public-Key and Hash-Based Cryptographic Schemes," Institute for Quantum Computing, University of Waterloo, February 7, https://arxiv.org/pdf/1902.02332.pdf.

[3] M. Roetteler, M. Naehrig, K.M. Svore, and K. Lauter, 2017, "Quantum Resource Estimates for Computing Elliptic Curve Discrete Logarithms," Cryptology ePrint Archive, https://eprint.iacr.org/2017/598.

gate error rate for physical qubits as a short-term metric and the effective number of logical error-corrected qubits of a system as a long-term metric to monitor the scaling progress of quantum computers.[4]

Disruptive

In the area of quantum cryptanalysis, a disruptive scientific advance would be experimental construction within the next decade of a quantum computer capable of threatening widely used public-key cryptography. This would likely involve the development of physical qubits with a sufficiently low error rate and the ability to build and connect these together at scale.[5] A similar disruptive effect might be achieved by an algorithmic improvement that threatens public-key cryptography without requiring an enormous number of error-corrected qubits.

The authors of the *Quantum Computing* report conclude that disruptive scientific advance in quantum computing is more likely if there is a virtuous cycle of research and development investment from industry. In such a cycle, initial scientific advances in quantum computing might attract large amounts of capital, which in turn could result in new rounds of significant scientific progress. An important advance that would enable such a virtuous cycle to fund the enormous engineering effort required would be the development of a currently unknown but compelling commercial application for noisy intermediate-scale quantum (NISQ) computers. Such a commercial application of NISQ computers would likely not be related to cryptography or security at all: machine learning or chemistry simulations are probably more lucrative and of broader interest. However, a virtuous cycle that defines a path to commercially relevant quantum computing at scale would likely bring about improvements speeding cryptographically relevant quantum computers as a side effect.

Last, there is the possibility of a scientific advance that conclusively rules out the possibility of scalable general purpose quantum computing under some set of conditions relevant to public-key cryptography. For example, there may be a proof of a lower bound on achievable error rates at levels incompatible with effective quantum computation. Such an advance would eliminate the potential threat to encryption from quantum computers.

Classical Cryptanalysis

Improvements to classical cryptanalysis will continue to be a factor in choice of security parameters and cryptographic algorithms going forward. History can be a guide for what these improvements and transitions may look like. There are several recent examples of cryptographic algorithm transitions that took place owing to improvements in classical cryptanalysis. These include the moves from the MD5 hash function to SHA-1 and from SHA-1 to SHA-2 after a progression of weaknesses and finally practical attacks were found on the MD5 and SHA-1 algorithms and the move from 1024-bit RSA to 2048-bit RSA (or larger) owing to expected improvements from Moore's law bringing 1024-bit factoring within feasible range.

Predictable

The current record for factoring the product of large prime numbers, set in 2020, is an attack on an 829-bit RSA modulus that took 2,700 core-years on a few academic clusters.[6] A back-of-the-envelope extrapolation from this running time suggests that a 1024-bit factorization would require around 500,000 core-years without any

---

[4] From NASEM, 2019, *Quantum Computing: Progress and Prospects* (Washington, DC: The National Academies Press, https://doi.org/10.17226/25196):

Primitive Boolean operations, implemented through digital logic gates, are the building blocks of contemporary computation. While quantum computers have bit-like structures (called "qubits") and gates, they behave very differently from classical bits and digital gates ... unlike classical gates, the quantum gates have no noise margin (input errors are passed directly to output of the gate) but their digital nature provides a means to recover from this critical drawback.

[5] X.-C. Yao, T.-X. Wang, H.-Z. Chen, W.-B. Gao, A.G. Fowler, R. Raussendorf, Z.-B. Chen, et al., 2012, Experimental demonstration of topological error correction, *Nature* 482(7386):489–494, https://doi.org/10.1038/nature10770.

[6] F. Boudot, P. Gaudry, A. Guillevic, N. Heninger, E. Thomé, and P. Zimmermann, 2020, "Comparing the Difficulty of Factoring and Discrete Logarithm: A 240-Digit Experiment," Cryptology ePrint Archive, Report 2020/697, https://eprint.iacr.org/2020/697.

further algorithmic or implementation improvements. The largest supercomputers in the world currently have millions of cores. The public literature has suggested that 1024-bit factorization has been within reach for large organizations since 2010 at the latest. A public demonstration of 1024-bit factoring or computation of a prime-field discrete logarithm (believed to be a few times harder than factoring) would not be surprising: the main difficulty would be convincing an organization with sufficient computing power that this would be a good use of hardware. Progress on the elliptic curve discrete logarithm problem—the basis for the elliptic curve variants Elliptic Curve Diffie-Hellman (ECDH) and Elliptic Curve Digital Signature Algorithm (ECDSA)—has mostly consisted of taking advantage of hardware improvements; records within the past decade have ranged from 113-bit to 118-bit curves; in principle, a few more bits above that would be feasible.[7]

Disruptive

Even a modest-seeming algorithmic improvement to asymmetric cryptanalysis algorithms could have a potentially high impact, because the key sizes in current use are based upon the running times of the current best attacks. The best cryptanalytic algorithms for RSA and Diffie-Hellman are the "number field sieve" algorithms for factoring and discrete logarithm for large characteristic finite fields, respectively. In 2013, a series of breakthroughs for a closely related algorithm, the function field sieve for discrete logarithm in small characteristic finite fields, produced the first algorithmic improvements for this problem after more than two decades and a series of impressive new computational records.[8] These improvements did not turn out to threaten widely deployed cryptography, but even a much more modest improvement in this direction for the general number field sieve would have the potential to threaten the 2048-bit RSA and Diffie-Hellman key sizes in current use and would require either increasing the key sizes or replacing with schemes based on alternative cryptographic assumptions.

An improvement in the algorithms that are the best classical cryptanalysis of elliptic curve discrete logarithm would be similarly impactful: the security of 256-bit curves could be threatened by any algorithm that improves upon the current algorithms for elliptic curves. Such an improvement would be more surprising than an improvement for factoring or finite-field discrete logarithm given the existing lack of progress in this area.

Based on current understanding, the classical improvements that jeopardize the security of major standardized symmetric cryptographic algorithms (Advanced Encryption Standard [AES], the SHA-2 and SHA-3 hash functions) seem much less within the realm of possibility. In particular, AES has a comfortable security margin against all known families of attacks;[9] the discovery of an entirely new and practical family of attacks that applies to AES would be both disruptive and surprising. (This is in contrast to the situation with factoring and discrete log mentioned above, in which usable key sizes have little security margin against any improvement in a known class of attacks.) However, flaws would still be expected to be found in other symmetric algorithms used in more niche applications, such as proprietary or low-power algorithms, which typically receive less public scrutiny or must accommodate challenging constraints.

Most of the scientific expertise in the open research community in cryptanalysis is outside of the United States, primarily in Europe, so any improvements would be expected to come from international researchers. While the Internet has enabled broader dissemination of knowledge about cryptography and security, the high-level research training and careers required to understand and contribute to the state of the art require real investment in research infrastructure from government and industry.

---

[7] D.J. Bernstein, S. Engels, T. Lange, R. Niederhagen, C. Paar, P. Schwabe, and R. Zimmermann, 2016, "Faster Elliptic-Curve Discrete Logarithms on FPGAs," Cryptology ePrint Archive, Report 2016/382, https://eprint.iacr.org/2016/382.

[8] R. Barbulescu, P. Razvan, P. Gaudry, A. Joux, and E. Thomé. "A heuristic quasi-polynomial algorithm for discrete logarithm in finite fields of small characteristic." In *Annual International Conference on the Theory and Applications of Cryptographic Techniques*, pp. 1–16. Springer, Berlin, Heidelberg, 2014. https://link.springer.com/chapter/10.1007/978-3-642-55220-5_1.

[9] See A. Bogdanov, D. Khovratovich, and C. Rechberger, 2011, "Biclique Cryptanalysis of the Full AES," in *Advances in Cryptology—ASIACRYPT 2011* (D.H. Lee and X. Wang, eds.), ASIACRYPT 2011, Lecture Notes in Computer Science, Volume 7073, Berlin, Heidelberg: Springer, https://doi.org/10.1007/978-3-642-25385-0_19, and P. Derbez, P.-A. Fouque, and J. Jean, 2012, "Improved Key Recovery Attacks on Reduced-Round AES in the Single-Key Setting," Cryptology ePrint Archive 2012/477, https://eprint.iacr.org/2012/477.pdf.

FINDING 4.1: Most of the current public scientific expertise in algorithm design, cryptanalysis, and other areas of applied cryptography is outside the United States, largely in Europe. In contrast, within the United States, cryptography is taught as an area of theoretical computer science. The specific areas of expertise necessary to guide and facilitate the transition to post-quantum cryptography are relatively new and will require a more robust educational pipeline to train new talent.[10] Public research investment, through the National Science Foundation and other organizations, would encourage this process, while strict U.S. export control regulations have historically discouraged talent from locating in the United States.

FINDING 4.2: An improvement in asymmetric cryptanalysis algorithms could have a significant effect on the security of public key encryption algorithms that are in wide use today. Such an improvement would enable more efficient attacks on encrypted information using conventional computers rather than requiring the construction of a quantum computer. Furthermore, it could potentially be exploited in secret and with little or no advance notice.

Post-Quantum Algorithms

The current candidates for post-quantum public key encryption and signature algorithms (lattices, structured lattices, supersingular isogeny Diffie-Hellman, hash-based signatures, etc.) all have downsides in the form of larger keys, ciphertext/signature size, and/or longer computation time than elliptic curve cryptography. Furthermore, the relative quantum resistance of each of the algorithm candidates under consideration for standardization is an area of active research. The cryptographic community continues to study the security assumptions underlying the candidates. The more public analysis and scrutiny a candidate receives, the more confidence the community gains in its presumed resistance to attack by a quantum-aided adversary. (See Chapter 2 for more details.)

Predictable

At a predictable rate of scientific advances, within the decade the committee would expect to see increased confidence in the resistance to classical and quantum cryptanalysis of these schemes, and better optimizations and instruction-level support in major central processing units (CPUs) to improve the performance of these algorithms, as well as support in transport layer security (TLS) and other major protocols. Some weaknesses in some special case parameter settings for these algorithms might be found, but it is reasonable to expect that the community will agree on a suite of implementable, post-quantum secure algorithms.

Disruptive

In this realm, the committee can foresee both constructive and destructive scientific advances. On the constructive side, a new quantum-secure construction might be found that does not have any key length or computational downsides compared to elliptic curve cryptography: this could greatly increase the speed of adoption. On the destructive side, an efficient quantum or classical algorithm for solving the shortest vector problem in algebraically structured lattices (or another post-quantum construction that the community currently has a lot of confidence in) would seriously disrupt the standardization and deployment process. The committee views significant cryptanalytic attacks against hash-based signatures as very unlikely because the security of these constructions relies on several hardness assumptions of the underlying hash functions. (See the related finding in Chapter 2.)

---

[10] To understand the cryptographic landscape, one must receive a Ph.D. in cryptography with at least 3–5 years of highly specialized training in graduate school. Even though the information is freely available on the Internet, the sheer volume of information and high degree of specialization means that without hands-on advising, it is nearly impossible to learn the skillset necessary to become proficient in cryptography.

Computing on Encrypted Data

There are several cryptographic technologies that accomplish tasks related to the general goal of computing on encrypted data in an untrusted environment without decryption. These include multi-party computation (MPC), searchable encryption, private set intersection, and partially and fully homomorphic encryption. (These were introduced in Chapter 2.) Currently, the use of these technologies requires the support of expert cryptographers.

Predictable

There have already been several deployed and widely used applications (and many proofs-of-concept) of MPC and searchable encryption. The predictable future, driven especially by system implementers' need to protect users' privacy, is for more MPC applications to be deployed, but without breakthroughs each will continue to require specialized expertise. Fully homomorphic encryption is less efficient than MPC. The other techniques will also see some applications but will remain niche technologies. There is substantial uncertainty in the use of zero knowledge (ZK) tied to blockchain techniques, as it is hard to foresee how the tension between blockchains and trusted third parties will evolve.

Disruptive

Advances in tooling, or in technology that makes it practical for non-experts to deploy new multiparty applications, would result in wider use of this form of confidential computing. On the other hand, if these techniques have to be made quantum-resistant their spread will be slowed. Algorithmic breakthroughs could make any of these techniques more efficient, and so more widely used. Regulations could limit or motivate the development and use of computation on encrypted data.

Anonymous Communication and Metadata Protection

In many existing communication systems, encryption is used to protect only the data considered to be the "content" of communications, while leaving metadata about sender and receiver, time, length, or other properties unencrypted. For example, Hypertext Transfer Protocol Secure (HTTPS) encrypts the contents of websites, but a passive observer of the network traffic can see the network IP (Internet protocol) addresses of the sender and receiver, the timing, length, and order of network requests, and other metadata that can reveal significant information about the likely encrypted content.

There are a number of cryptographic tools and ideas that can be used to build privacy-enhancing technologies by allowing systems to function while concealing information from some or all parties. For example, the Tor network uses cryptography to route network connections through volunteers in order to keep an eavesdropper from linking the connection's origin and destination network address together. The Apple/Google COVID-19 exposure notification system uses cryptographic ideas to maintain the participants' privacy while still reporting their contact status.[11,12]

In addition to the concepts of encryption, MPC, and FHE discussed above, a number of advanced cryptographic constructions that are not currently widely used can achieve more advanced privacy goals; these include zero-knowledge proof and its variants, function secret sharing, Oblivious RAM, and Private Information Retrieval. These ideas are discussed in Chapter 2.[13,14]

---

[11] There is some leakage in the Apple/Google COVID-19 solution, which is addressed in additional proposals.

[12] Apple, 2020, "Apple and Google Partner on COVID-19 Contact Tracing Technology," Newsroom, April 10, https://www.apple.com/newsroom/archive.

[13] K. Yang, P. Sarkar, C. Weng, and X. Wang, 2021, "Quicksilver: Efficient and Affordable Zero-Knowledge Proofs for Circuits and Polynomials over Any Field," Cryptology ePrint Archive, https://eprint.iacr.org/2021/076.

[14] N. Franzese, J. Katz, S. Lu, R. Ostrovsky, X. Wang, and C. Weng, 2021, "Constant-Overhead Zero-Knowledge for RAM Programs," Cryptology ePrint Archive, https://eprint.iacr.org/2021/979.

Predictable

Recent years have brought real-world deployments of these ideas, such as the use of ZK to build cryptographically untraceable anonymous transactions in the cryptocurrency Zcash, and deployment of MPC by Google and Apple for applications like computation on sensitive data and making assertions about private data without disclosing the underlying data.[15,16] Predictable scientific advances in these areas would include efficiency improvements across the board. A predictable course of development would also see companies deploying more sophisticated cryptographic schemes to protect aspects of privacy and verifiability for their users and data. As these applications broaden, this will drive a cycle of increased interest and investment in efficient cryptographic privacy-preserving techniques and implementations in both industry and government applications.[17] A predictable counter-trend is that relying parties will struggle to understand the protocols or be (perhaps rightly) skeptical of the security of implementations, limiting the benefits that these technologies can ultimately provide.

Disruptive

Disruptive advances in this area may come from transitioning technologies into practice. For example, a completely anonymous low-latency network communication mechanism that produces no metadata and sees wide adoption would likely be disruptive. A second example of a disruptive development might be government initiatives to use ideas from privacy-enhancing technologies to build more sophisticated or privacy-preserving mechanisms for law enforcement search or tracking of encrypted communications and data. As a positive development, such initiatives could have the potential to provide more transparency or accountability for government agencies if implementations are rigorous and transparent. They may also spur development and refinement of privacy-enhancing technologies. On the other hand, such deployments risk public backlash, misunderstandings over the technical guarantees provided by these technologies, misuse, or loss of trust owing to implementations being vulnerable to the same types of vulnerabilities and bugs common to any implementation.

Lightweight Cryptography

The AES cipher is near ubiquitous for symmetric encryption, and no flaws are expected to be found that might call its security into question. It was designed to be well suited to hardware implementation, and many microprocessors have high-speed AES hardware support.[18,19] However, there have been calls to develop and standardize new "lightweight" symmetric encryption algorithms that would be suitable for extremely small and low-resource sensors expected to be part of the coming IoT, and the National Institute of Standards and Technology (NIST) is currently in the process of selecting lightweight cryptographic algorithms for standardization.[20]

Predictable

There are already a number of candidate lightweight cryptographic algorithms proposed to NIST for standardization. Predictable scientific advances in this area would include successful conclusion of the NIST activity and

---

[15] E. Ben Sasson, A. Chiesa, C. Garman, M. Green, I. Miers, E. Tromer, and M. Virza, 2014, "Zerocash: Decentralized Anonymous Payments from Bitcoin," *2014 IEEE Symposium on Security and Privacy*, https://doi.org/10.1109/sp.2014.36.

[16] M. Ion et al., 2020, "On Deploying Secure Computing: Private Intersection-Sum-with-Cardinality," pp. 370–389 in 2020 IEEE European Symposium on Security and Privacy (EuroS&P), https://doi.org/10.1109/EuroSP48549.2020.00031.

[17] For example, see the proposed "H.R.4479—Student Right to Know Before You Go Act," which would use MPC to compute information about college student outcomes.

[18] Wikipedia, "Trusted Platform Module," Wikimedia Foundation, https://en.wikipedia.org/wiki/Trusted_Platform_Module, accessed October 14, 2021.

[19] Intel, "Intel® Software Guard Extensions," https://software.intel.com/content/www/us/en/develop/topics/software-guard-extensions.html, accessed October 12, 2021.

[20] NIST, "Lightweight Cryptography," Computer Security Resource Center, Information Technology Laboratory, https://csrc.nist.gov/Projects/Lightweight-Cryptography, accessed October 15, 2021.

publication of a Federal Information Processing Standard (FIPS) for lightweight cryptography along with guidelines for appropriate use scenarios, followed by some deployment in real-world systems with those characteristics.

Disruptive

A disruptive scientific advance in this area might include a catastrophic break of an algorithm after its standardization and large-scale deployment.

Standards for Cryptography

A number of different international and national organizations publish standards and recommendations for cryptographic algorithms and protocols. Examples of organizations publishing influential cryptographic standards include NIST, which publishes algorithm recommendations that vendors must follow in order to be FIPS certified and sell their products to the U.S. government, the Internet Engineering Task Force (IETF), which publishes standards for network protocols including TLS, SSH, S/MIME, IPsec, and many others, and the International Organization for Standardization (ISO), which publishes international standards.

These organizations use different processes for creating and approving standards. For several major cryptographic algorithm recommendations (AES, SHA-3, and the post-quantum cryptography algorithms), NIST has run open international competitions in which researchers submit their algorithm designs and those designs go through several rounds of vetting, including attempts at cryptanalysis, by the open research community and NIST before NIST chooses and publishes the final candidates. NIST also publishes recommendations written through an internal process that involves workshops and public comment periods.

Protocols standardized and published by the IETF are the result of open processes and the work of individuals within the IETF's various working groups (WGs). Anyone can join an IETF WG and propose or contribute to the development of a draft standard, and only individuals can be members of a WG. Although the character of individual WGs vary, common processes are used across the IETF to develop their standards. The resulting Requests for Comments (RFCs) document most of the protocols upon which the Internet operates today. Recently, the IETF has begun collaborating more closely with academia on the development of new security standards; the TLS 1.3 RFC is an example of a particularly open collaboration among academic and industry contributors.

ISO committees are composed of representatives from different countries' national standards bodies such as the American National Standards Institute (ANSI) in the United States. Industry, academic, and individual contributions to ISO standards and the participation of subject matter experts are channeled through the national bodies.

In addition to the organizations listed above, sometimes cryptographic standards are developed by industry groups or consortia as part of an effort to standardize new technology. For example, both the 3rd Generation Partnership Project (3GPP), which developed mobile telecommunications standards such as Global System for Mobile Communications (GSM) and LTE, and the Bluetooth Special Interest Group, which developed the various Bluetooth short-range wireless specifications, designed and specified their own encryption algorithms as part of their specification suites.

There are multiple technical and non-technical factors that drive the development of new standards and recommendations. Some of these include desire for new functionality, performance, desire to obtain patent royalties or impose licensing restrictions, desire for backdoors or other mechanisms that could enable government access to encrypted data, desire to advantage national industries and economies, and nations' preference for relying on their own standards and technologies. The costs in time and effort of participating in standards bodies can limit participation to individuals or organizations who seek a commercial or reputational benefit from the integration of a particular idea or technology in a standard or from blocking an idea or technology.

Predictable

NIST is expected to produce post-quantum encryption and digital signature algorithm recommendations starting in early 2022 and continuing for the next few years with multiple waves of selections. All of the final candidates were produced by researchers working in the open community. Owing to the open process and NIST's history of past successful competitions (e.g., AES, SHA-3), the predictable path of development and deployment is for NIST's selections to be adopted into updated security standards and protocols over the next 5 to 10 years.

However, because the structure of these algorithms is new, the committee also expects continued research and development of newer post-quantum algorithms and variants of NIST's selections. A predictable path of development for these variants, if they offer measurable performance or security advances over NIST's selections, would see them potentially adopted by major implementations with sufficient cryptographic agility in parallel with NIST's selections. Eventually, these variants might themselves be included in updates to NIST's standards.

Disruptive

A potential disruptive development might involve the U.S. government's dilemma upon the discovery of a weakness in an algorithm that had been proposed to be standardized. Although an Executive Branch policy process known as the Vulnerabilities Equities Process exists to enable a decision either to disclose fully the details of the weakness or to "retain the vulnerability" for the Intelligence Community's possible use, neither option might be attractive. The government might have to disclose classified details to convince a mistrusting, skeptical public, influenced by the memories of the Dual Elliptic Curve incident detailed in Chapter 2, that the algorithm was indeed deficient; but if that were unacceptable, it would withhold such information, thus allowing use for public and commercial purposes of the weak algorithm.

Other disruptive developments might include wildly divergent algorithm recommendations coming from Chinese or European government standards bodies, or the standardization at a national level of an algorithm suspected to be backdoored by a government agency that large technology companies are forced to implement in order to sell products in that country.

Program Obfuscation

There has been considerable interest in cryptographic methods for program obfuscation, but it remains an open question whether an efficient general approach can be built for practical applications, despite a considerable output of research publications on the topic.[21]

Predictable

Cryptographic software obfuscation may never be feasible even in principle. An alternative to software-based cryptographic program obfuscation is to use hardware security mechanisms to protect the execution of sensitive programs. Examples of such hardware mechanisms in use today include Apple's Secure Enclave technology and trusted execution environments like Intel's SGX.[22,23] The current state of the art is that there is a cat-and-mouse game between security researchers and hardware developers to discover methods to circumvent these mechanisms and patch the flaws.[24] The predictable course of development in this area is that secure hardware technology will steadily improve over time through this process, and hardware-enforced security will become even more practical and widely used. See the discussion of tamper-resistant hardware in the Systems Driver section below for more details.

---

[21] S. Garg, C. Gentry, S. Halevi, M. Raykova, A. Sahai, and B. Waters, 2016, Candidate indistinguishability obfuscation and functional encryption for all circuits, *SIAM Journal on Computing* 45(3):882–929, https://doi.org/10.1137/14095772x.

[22] Apple, "Secure Enclave," Support, https://support.apple.com/en-gb/guide/security/sec59b0b31ff/web, accessed October 12, 2021.

[23] Intel, "Intel® Software Guard Extensions," https://software.intel.com/content/www/us/en/develop/topics/software-guard-extensions.html, accessed October 12, 2021.

[24] S. Van Schalik, A. Kwong, D. Genkin, and Y. Yarom, "SGAxe: How SGX Fails in Practice," https://sgaxe.com/files/SGAxe.pdf, accessed October 13, 2021.

Disruptive

There are recent scientific results showing that some types of program obfuscation can be built with security based on plausible hardness assumptions. However, even if the security of these proposed schemes and hardness assumptions holds up to scrutiny, there is a long road to practicality: the initial proposals are too slow for any practical application. A fundamental breakthrough or series of breakthroughs would be needed to make a provably secure program obfuscation scheme based on well-accepted hardness assumptions practical; such a fortuitous series of developments happening within 10 to 20 years would be unexpected.

Cryptocurrencies

Predictable

The level of interest and funding around cryptocurrencies has already begun to spur scientific development of related technologies, including consensus protocols, short zero-knowledge proofs, advanced digital signatures, time-lock puzzles, cryptographic protocols, and verifiable computation. Cryptocurrencies face some major barriers to more widespread adoption, notably the possibility of widespread government regulation to limit untraceable transactions. Other barriers may include environmental impact for proof-of-work based schemes, transaction delay, transaction traceability, and security considerations including fraud and irreversible transactions. The scientific tools necessary to solve the computational problems are getting better understood.

There is considerable hype around applications of blockchains, but the so-called permissionless (typically proof-of-work or proof-of-stake based) blockchains that are used to back the most common cryptocurrencies, as well as the underlying data structure of a blockchain itself, seem poorly suited to many applications. Permissionless blockchains involve serious additional barriers to broad adoption owing to environmental impact, security challenges, and inefficiency. In a predictable sequence of events, the use of such permissionless blockchains would largely be a symbolic and minor detail in what are effectively centralized applications. Developers of distributed systems may also continue to use the word "blockchain" to apply to the more sophisticated consensus algorithms, multiparty computation, and more advanced distributed data structures that would be a better match for bona fide distributed applications.

A predictable course of development over the next 10 to 20 years would see cryptocurrencies and related technologies gain somewhat broader adoption and continue to serve as inspiration, funding source, and proving ground for an increasing variety of advanced cryptographic techniques. It is also predictable that cryptocurrencies currently based on classical cryptography (e.g., ECDSA) will begin to migrate to post-quantum algorithms along with other industries on similar time scales. Predictable developments could also include the rapid migration from the existing proof-of-work scheme (e.g., Bitcoin mining uses proof of work) to alternative schemes (such as proof of stake; Ethereum plans to transition within this year) that have much lower environmental costs and impacts.

Disruptive

It is possible that disruptions could emerge toward growth or contraction with respect to cryptocurrencies. Disruptive growth might see sufficiently broad deployment that one or more cryptocurrencies become the de facto fiat currency in large parts of the world. Disruptive contraction could come largely from the policy arena discussed below under "Society and Governance" rather than technical developments, but possible developments are listed here to provide a complete picture. Disruptive contractions could occur if cryptocurrencies featuring privacy and anonymity wither and eventually die under a barrage of strict government regulations related to money laundering, sanctions evasion, terrorist financing, fraud, cyberattacks, and similar concerns. Cryptocurrencies—Bitcoin in particular—facilitated the monetization of international ransomware attacks, and continued attacks could lead to significant regulatory changes. The investment bubble surrounding cryptocurrencies is large enough that a catastrophic crash could have measurable economic impact; such a crash could result in loss of interest, excitement, and funding in the general area.

Applications of Cryptography

Cryptocurrencies are an application of cryptography that futurists envisioned in the 1980s, and became a reality decades later with the first release of Bitcoin in 2009. Cryptocurrencies began to drive cryptographic and systems advances when they took off during the 2010s, as well as enabling additional chaos by facilitating the evolution of ransomware. There are numerous other potential applications of cryptography that could be developed to the point of practicality, including privacy-preserving collection and search methods for law enforcement and intelligence, cryptographically verified supply chains, cryptographically verified voting receipts,[25] or using the cryptographic techniques described above to allow privacy-preserving data sharing in business contexts.[26]

Predictable

Other applications of cryptography that were predicted by the "cypherpunks" and other enthusiasts and have since seen real-world use include anonymized web browsing, hidden services and the "dark web," dark markets, prediction markets, steganographic techniques, and ubiquitous end-to-end encrypted messaging.[27] A predictable rate of scientific advances would see incremental improvements to the efficiency and usability of these technologies by non-experts that would enable more widespread deployment. These new applications can have powerful effects, often simultaneously for good and bad, with controversy attached. This is already true for secure communications and cryptocurrencies; other applications of cryptographic technologies would likely also be seen to be used for both good and bad.

Disruptive

A disruptive advance in this category would be an idea like Bitcoin that goes from a not fully realized intention to proof of concept to mainstream at a rapid pace and spurs a cycle of investment, further scientific advances, and ultimately societal impact.

Quantum Key Distribution

Quantum key distribution (QKD) systems leverage properties of light quanta to enable parties to negotiate a secret key that an eavesdropper cannot determine except through an exceptional stroke of luck. (Despite sharing the term "quantum" and sometimes being referred to as "quantum cryptography," QKD is entirely unrelated to quantum computing or post-quantum cryptography.) Current QKD designs perform only key agreement; authentication would need to be done using standard cryptographic algorithms, although someday long-term quantum memories could perhaps support authentication via unclonable shared, secret keys. Although QKD systems are often described as implementing one-time pads, at present they cannot generate key material fast enough to protect high-speed traffic; hence they may be used to generate key material for classic algorithms such as AES, enabling changes of the symmetric key very often (e.g., many times per second).

The security properties of real QKD systems are somewhat murky at present, and because they require transmission and reception of exceptionally dim light ("single photons") they need specialized optical equipment and an unencumbered optical channel through fiber or the atmosphere. Although such systems may have niche applications, they are unlikely to become pervasive within the next 20 years.

---

[25] An example of one such system is the ElectionGuard end-to-end verifiable election technology by Microsoft (T. Burt, 2019, "Protecting Democratic Elections Through Secure, Verifiable Voting," *Microsoft on the Issues* (blog), May 6, https://blogs.microsoft.com/on-the-issues/2019/05/06/protecting-democratic-elections-through-secure-verifiable-voting). Recently, Hart InterCivic (one of the "big 3" voting equipment manufacturers in the United States) announced that it would start incorporating ElectionGuard into its systems (Microsoft News Center, 2021, "Hart and Microsoft Announce Partnership to Make Elections More Secure, Verifiable," June 3, https://news.microsoft.com/2021/06/03/hart-and-microsoft-announce-partnership-to-make-elections-more-secure-verifiable).

[26] Microsoft News Center, 2021, "Hart and Microsoft Announce Partnership to Make Elections More Secure, Verifiable," June 3, https://news.microsoft.com/2021/06/03/hart-and-microsoft-announce-partnership-to-make-elections-more-secure-verifiable.

[27] Cryptoanarchy.wiki, "Cypherpunks Mailing List Archive," https://mailing-list-archive.cryptoanarchy.wiki, accessed October 15, 2021.

Predictable

A predictable course of advancement might see existing proof-of-concept networks in China, Japan, Europe, and perhaps the United States grow through the addition of further links over hundreds of kilometers. Further demonstrations of satellite QKD may happen, and perhaps small proof-of-concept satellite networks. Commercial systems may find niche markets.

Disruptive

Convincing demonstrations of QKD implementations that provide high security in practice may lead to adoption in some specialized systems. Real-world demonstrations of quantum repeaters and quantum networks would be a significant technical achievement and could enable somewhat broader use. However, even a major technical advance would not be expected to have an overall disruptive impact owing to the underlying limitations of QKD.

Artificial Intelligence and Machine Learning

The fields of artificial intelligence (AI) and machine learning have made significant progress in recent years and are receiving an enormous amount of attention in the research community as well as growing real-world use. There are multiple ways that cryptography interacts with machine learning. A major one is the use of encryption to protect the privacy of potentially sensitive data used to train machine learning models. Cryptographic tools for computing on encrypted data like homomorphic encryption and multiparty computation can be used to train machine learning models on encrypted data, or to encrypt the models themselves.[28] In applied security research, there has been significant recent work on adversarial machine learning, including attacks that fool machine learning models into making incorrect choices, or invert machine learning models to learn sensitive information about training data. Machine learning has been used in applied security applications, such as in anomaly detection for network data. Current applications to core areas of cryptography are more limited. Machine learning has been used to analyze side-channel traces, and a handful of works have explored using machine learning techniques to design or analyze ciphers.[29]

Predictable

A predictable set of advances would see the proof-of-concept demonstrations of privacy-preserving machine learning using cryptography to protect training data or the model becoming more sophisticated. It is unlikely but not impossible that machine learning would lead to significant new attacks on deployed cryptography. There may be academic proof-of-concept works using machine learning to optimize fast implementations, discover new side channels, or improve upon existing work learning differential paths or other areas of classical cryptanalysis.

There are also numerous cryptography-adjacent applications of machine learning to security more generally. For example, analyzing metadata, analyzing data decrypted owing to flawed cryptography or implementations, correlating unencrypted data sources with encrypted sources, optimizing computational clusters for cryptanalysis or data analysis, and so on. A predictable set of advances would see all of these applications grow more sophisticated and may also inspire additional work on cryptographic techniques to protect against these techniques.

Disruptive

A disruptive advance might be the development of an entirely new field of cryptanalysis based on machine learning that is able to find subtle flaws and biases in published cryptographic algorithms much more effectively

---

[28] C. Juvekar, V. Vaikuntanathan, and A. Chandrakasan, 2018, "Gazelle: A Low Latency Framework for Secure Neural Network Inference," arXiv.org, January 16, https://arxiv.org/abs/1801.05507.

[29] A. Gohr, S. Jacob, and W. Schindler, 2019, "CHES 2018 Side Channel Contest CTF: Solution of AES Challenges," Bundesamt für Sicherheit in der Informationstechnik (BSI), https://eprint.iacr.org/2019/094.pdf.

than the statistical/computational methods or human-driven mathematical analysis tools that are used to analyze algorithms today. A significant flaw or bias discovered in a deployed algorithm that could not be found by human analysis would be disruptive to the cryptographic community, because it would mean that a new set of analysis tools would need to be added to the analyst's toolkit before a new algorithm could be trusted.

Dramatic advances in the ability of AI systems to write or analyze code, or to test systems, could make a mature systems future much more likely, by making it economically feasible to generate new and more reliable systems, or to find the flaws in existing systems.

SOCIETY AND GOVERNANCE DRIVER: GLOBALIZATION VERSUS FRAGMENTATION

Figure 4.2 defines what is meant by the Society and Governance driver, and how a future would look if it existed at either extreme.

How countries and their citizens will perceive risks and benefits of encryption, and how in response they will choose to protect cultural values and develop local laws and policies, will obviously affect the future of encryption in both direct and diffuse ways. Although there are many facets to the Society and Governance driver, the committee believes that a critical differentiating feature will be the extent to which this driver leads to global convergence—resulting in countries around the globe taking similar approaches to the relevant issues—or fragment along national, cultural, or other geopolitical lines.

As a threshold matter, the Society and Governance driver presents three analytic challenges to a greater degree than some of the other, more technical drivers the committee considered:

Driver Details: Society & Governance

How nation states and individuals perceive risks and benefits of encryption and have designed standards, laws, and policies to address them

Below are example polar extreme characteristics of the Society and Governance driver:

Fragmented ⟷ **Global**

Fragmented	Global
Autocratic and authoritarian governments control internet content for political reasons, demanding local data storage, limiting operations of global tech companies, and have access to digital communications for domestic internal security purposes.	Technological development generally, and the internet in particular, promote globalized standards and approaches; the growth of multinational corporations, including "Big Tech," leads to worldwide solutions, often with self-reinforcing elements; and governments cooperate more to address global economic, security and environmental challenges
Nationalist regulation results in significant limits on global interoperability of hardware, software, and data.	Efforts to address cybersecurity, online hate speech and disinformation are emulated globally, with a generally benign effect
Mistrust of government and institutional deligitimization leads to extensive end-to-end encryption for communications; corporations are trusted more and have more access to information.	Governments and the IC rebuild trust with societies and are granted more access to information, so that law enforcement and IC generally have access to encrypted information

FIGURE 4.2 Visual representation of space defined by Society and Governance driver.

1. The Society and Governance driver tends to be inherently subjective, ambivalent, diffuse, and non-quantitative.
2. This driver is often interrelated and does not operate in a mutually exclusive manner—for example, even in an extreme case of globalization, powerful countervailing trends for fragmentation will still be present. The very factor propelling one outcome may readily trigger the opposite result. (For example, disintegration of a country's political structures might lead to a weak national government with local powers predominating and competing, but might also lead to a stronger, perhaps authoritarian national regime.)
3. Even when a particular driver manifests itself strongly, the consequence for intelligence collection or defense might not necessarily be correspondingly clear. (For example, a globally common approach to technology regulation might mean that the Intelligence Community would need to devote fewer resources to solving access problems, because one success could presumably be easily replicated elsewhere; but it could also mean that all targets are similarly problematic and by definition there would be fewer weaker individual targets.) In other words, the relative strength of a particular aspect of the Society and Governance driver may not translate directly into a particular result for encryption or an unalloyed positive or negative outcome for the Intelligence Community.

The committee arrived at the decision to use the Society and Governance driver for its scenario planning, and within that, to explore globalization versus fragmentation, prior to the publication of the *Global Trends 2040* report.[30] Nonetheless, the committee believes that report presciently outlines the major themes and trends warranting consideration.[31] Most significantly, it is worth noting that three out of the five themes expressly outlined in the report—fragmentation, disequilibrium, and contestation—all tend toward worldwide differentiation in a wide array of areas. Every one of the five scenarios examined by the report has strong, and in most cases, dominating elements of fragmentation and differentiation. Given the thorough discussion of broad societal drivers contained in Global Trends, the committee does not repeat that analysis, but instead supplements it here with a particular focus on how those drivers might manifest themselves in specific topics particularly relevant to encryption.

This section identifies and discusses what the committee believes to be the most salient drivers for globalization or fragmentation, which arise from a variety of sources. One dynamic, however, affects almost every part of the discussion: the rise of China as an economic, military, technological, and geopolitical power. Although China's development is going to have a profound effect on the drivers discussed below, and the velocity of that rise is uncertain (and possibly unequal in various sectors), it is not clear whether the force of an ever-more potent China, and the rest of the world's reaction, will automatically lead to globalization or fragmentation (although the latter seems more likely). Both globalization and fragmentation are discussed in depth below, but the committee nonetheless believed that the role of China in driving each of these two trends to be of such overriding potency that it warrants specific mention.

Last, the committee recognized that many of the factors tending toward globalization are offset by other factors pushing in the opposite direction even where both arise in the same context. The committee decided that it was better to organize this discussion by exploring the factors favoring globalization in one section and favoring fragmentation in another. Another option would have been to organize by subject area, and explore each area's effects toward globalization or fragmentation, but this would have led to a more cumbersome presentation. The committee appreciates that presenting the factors separately is artificial because it is often the case that forces for both globalization and fragmentation are present at the same time in a single situation but believes that this approach may be more useful for the reader. Given that many of the concepts favoring globalization are straightforward and easily apprehended, the committee's discussion starts with those concepts, and then separately considers fragmenting forces.

---

[30] National Intelligence Council, 2021, *Global Trends 2040: A More Contested World*, Office of the Director of National Intelligence, March, https://www.dni.gov/index.php/gt2040-home, accessed October 21, 2021.

[31] Appendix D summarizes relevant points from the *Global Trends 2040* report.

Factors Favoring Globalization

The Role of the Internet

The Internet—arguably the most transformational manifestation of the Digital Revolution—has directly fostered globalization, owing to its inherently open and interoperable nature. It has propelled the free and easy flow of information, ideas, and communications without regard to sovereign boundaries. Thus, everything from news to pop culture to products and services to political concepts (including democracy, populism, and authoritarianism) has the potential to reach a worldwide audience instantly. The resulting shared experience is itself a powerful cause of international commonality.

Many of the applications available via the Internet are expressly intended to operate across borders, with easy adaptation to multiple languages and services ranging from purchasing to messaging designed for multinational users. Although technically possible, it is difficult for a nation-state to filter Internet traffic selectively at its border, and the political, cultural, and economic downsides of blocking all or significant portions of Internet traffic for a country are substantial. As those cross-border factors are sustained, the Internet will continue to be a force for globalization.

More broadly, the fact that the Internet is a manifestation of global adoption of technical standards itself facilitates global approaches to electronic communications generally. Although the Internet was initially the product of U.S. national security research, its evolution and technical workings owe relatively little to direct government efforts, and far more to those of the private sector. Governmental involvement has consisted mostly of promoting global connectivity and interoperability—in the case of the latter, principally by supporting various international professional bodies that set technical standards, and by adopting the resulting standards in national regulation.

Moreover, many countries have not stopped encrypted communications at their borders—the Internet grew to its current scale without most national governments governing the technical workings of Internet traffic that transits, originates, or terminates in a particular nation. In addition, multiple telecommunication modes such as voice, video, data, and text messaging are highly interconnected, providing users additional channels for cross-border communication. By contrast, in the early days of telephony, countries imposed effective technical, geographical, and substantive limits—the communications of the Soviet Union and its Warsaw Pact allies, for instance, existed largely separate from communications networks of the United States and its allies.

Other Technological Developments

Although the Internet may be a particularly visible cause and effect of modern globalization, technological developments have also led to the globalization of trade in goods and services generally. Diverse sectors including financing, manufacturing, logistics and shipping rely on the ease and lower cost of cross-border communications (voice, video, and data). Many industry sectors have increasingly adopted common standards around the world. Although protectionist pressures exist in many settings, the broad trend has been for governments to support (or at least not oppose) expanding cross-border flow of goods and services.

In particular, supply chains for many hardware and electronics products have become highly global, especially since China entered the World Trade Organization in 2001. So, for example, owing to the Internet and global communications, as well as standardization, a product sold in the United States might be designed in Europe, manufactured in Asia with parts from various countries, shipped around the globe, and subsequently supported by a customer call center in India. Common standards in all these areas are likely to be more successful where there are unified approaches to communications (including encryption). Indeed, the adoption of English as the principal language of commerce, finance, and certain other endeavors (such as aviation) itself promotes worldwide commonality to a significant extent.

All of this globalization is principally the result of private sector endeavors; it is that sector, not the governmental one, that is mostly leading technological innovation because of economic motivation. As major corporations with worldwide operations dominate this innovation and associated delivery of products and services, this contributes powerfully to globalization, and is likely to have an associated effect on encryption issues. By contrast, in an era in which individual governments lead or substantially regulate technological development, there is a greater likelihood of nationalistic approaches to such development.

As a corollary to the evolution of the worldwide Internet, there have been powerful trends toward global approaches to software and online services. U.S.-based companies have achieved widespread adoption globally for important categories of software, such as operating systems from Microsoft, Apple, and Google and social media platforms. In addition to proprietary software, which is now sold worldwide, open-source software is and will be important for a wide range of uses, and open-source software by its nature can be created or adopted anywhere in the world. Another important globalizing force is cloud computing, in which large processing and storage providers such as Amazon, Microsoft, Google, Salesforce, SAP, and many others afford customers around the globe the ability to store and use data in uniform ways. With software and cloud computing so often operating across borders, the advantages of interoperability create additional reasons to adopt international, and perhaps global, approaches for encryption, including for data in transit across borders.

Increasing reliance on satellites for communications and processing, including for remote monitoring for purposes ranging from climate change to agricultural production, will also propel the globalizing trends. Depending on whether adequate bandwidth can be available and transmitted and on the size of receiving antennas and the economics of low Earth orbit satellite constellations, it is possible that widespread Internet availability from satellites could sidestep or at least complicate national regulation of Internet access and could consequently promote globalization. While significant portions of satellite communications (both satellite to satellite as well as Earth connections) are currently not encrypted, in the future more elements of such communications might be encrypted; given the nature of those communications, there would be strong incentives to have globally common schemes.

Political Factors

There are, of course, numerous political institutions—mostly an outgrowth of World War II and partially in response to the Cold War—that expressly promote global and regional cooperation in matters ranging from security to trade, public health, and transportation, and to the extent those institutions continue to be effective, globalization will tend to predominate. In particular, as it relates to encryption, the "Five Eyes" intelligence sharing arrangement of English-speaking nations (i.e., the United States, the United Kingdom, Canada, Australia, and New Zealand), and their corresponding coordination in law enforcement, are also drivers toward greater emphasis on common approaches to encryption and intelligence collection generally.

More broadly, as nations seek to deal with common problems, they have historically pursued coordinated approaches, at least where consistent with national interests. Thus, global problems such as climate change, international terrorism, nuclear proliferation, and control of outer space might all elicit global approaches in a wider range of areas. The national security mission of the Intelligence Community appears to be expanding beyond traditional political and military adversaries to a greater array of such threats and vulnerabilities. The resulting increased responsibility might be made less onerous by reliance not only on the Five Eyes relationship (which itself could expand) but also on potentially numerous other like-minded nations seeking to address these global problems.

Self-Reinforcing Elements

Last, virtually all of the factors noted above as propelling globalization have strong reinforcing elements that deepen and expand the trend. On a simple level, the more a particular technology or standard is widely adopted, the more difficult or expensive it is for "holdouts" to persist with alternatives, and for competing approaches to arise and take hold. To some extent, this characteristic is true even of ideas and societal approaches; the past few decades have seen expansion of legal rules that protect human rights, including the rights of women and historically marginalized groups. Such protections suggest the possibility of globalization of world public opinion on important matters.

Factors Favoring Fragmentation

Even a casual reader of the preceding paragraphs would be quick to point out that there frequently is a countervailing trend for every factor tending to globalization. The committee does not attempt to detail every such trend, but again notes that this inherent duality or ambivalence is present in most aspects of the Society and Governance driver.

Potential for Misuse

Perhaps this duality, with its latent potential for abuse, is an aspect of the human condition and the way societies organize—but it is thrown into sharp relief by technology itself, which lends itself to later malicious use or reveals risks not fully appreciated at the time the technology was introduced. For example, when radio became common around the world in the 1920s, many observers thought that it would lead to world peace and initially viewed the invention as purely a benefit. Yet within a matter of years, totalitarian regimes in the Soviet Union and Germany used radio for propaganda. The same pattern is of course playing out today with the advent of (mostly previously underappreciated) cybersecurity risks and online maliciousness, as well as disinformation on the Internet.

The result of this greater appreciation of the inherent risks and the potential for misuse of modern communications systems supported by the Internet is in many cases a cause for national or local regulation. This is especially the case where there is a perception that effective global safeguards either do not exist, or are not tailored to support the country's political, economic, or cultural concerns. At least at this point in the evolution of the Internet and given the current situation of global political competition and a rise in nationalism and populism, it seems that the forces arrayed against globalization and furthering national regulation are gathering strength. The extent of the recent different approaches taken by China and to a lesser extent even the European Union to a wide array of U.S. technology is illustrative of the strength of these forces.

Regulation

In their most benign manifestation, these anti-globalization forces take the form of national regulations to promote online competition, enhance cybersecurity, curtail hate speech, and protect citizens' data privacy. The result can be a multiplicity of complex regulatory schemes that vary from nation to nation, which offset the predisposition of the multinational corporations that control the major online platforms (such as marketplaces, search engines, social media, payment systems) to operate in a uniform manner around the world.

There might, however, be some natural limiting element to this apparently benign regulation, because governments will be somewhat hesitant to curtail commercial enterprises that perform a useful and desired service. Consequently, regulation might turn out to be a somewhat weaker fragmenting force. On the other hand, to the extent that U.S. companies dominate social media, marketplace and communications aspects of the Internet, there could be a political backlash against perceived U.S. hegemony, and that might add fuel to the fragmenting forces. Both China and the European Union have separately sought to regulate U.S. social media, hardware and device manufacturers and online service providers in ways that require conformity to local rather than global rules.

Factionalism Fomented by Authoritarian and Autocratic Governments

By contrast, in its more nefarious form, the impetus for national control is a far more potent cause of fragmentation. The efforts on the part of regimes such as Russia, China, Iran, and Turkey to control content on social media platforms and to impose email censorship (along with extensive user surveillance) are the most visible manifestation of governmental pushback against the open and unregulated nature of the Internet.[32] To the extent that this trend intensifies, it could entail the end of the World Wide Web, and the development of the "splinternet"—with individual countries or blocs of like-minded countries imposing substantive content requirements enabled by technological distinctions at national levels. This could include, for example, prohibiting or regulating virtual

---

[32] Freedom House, 2021, "New Report: Global Battle Over Internet Regulation Has Major Implications for Human Rights," press release, September 21, https://freedomhouse.org/article/new-report-global-battle-over-internet-regulation-has-major-implications-human-rights.

private networks (VPNs), banning end-to-end encryption (so as to permit government surveillance), or mandating a variety of governmental access to otherwise encrypted communications (perhaps through required turnover of encryption keys to authorities or insisting on the use of specified encryption schemes).

In the case of China in particular, it can be difficult to isolate the relative strengths of the various drivers that propel development of their own unique systems, including nationalism, the push for use of Mandarin throughout the nation, the strategic and economic goals of the Chinese Communist Party for China's technological superiority, and the sheer mass of the country in terms of both geographic size and population. These are all factors that cause China to stand apart from the rest of the world. This is seen most readily with social media and online payment systems, where uniquely Chinese systems are extraordinarily pervasive within that country (e.g., WeChat has more than 1.2 billion users in China), and yet are not significant elsewhere (except among the Chinese diaspora). In any case, these drivers in one form or another are precipitating a fragmentation involving a fifth of the world's population.

Indeed, to the extent China remains a rising, if not yet dominant, power, its sheer presence on the world stage is itself a fragmenting element. Seeking to control the internal communications of its own population, China is in the process of developing its own approach to communications, separate from the U.S.-dominated international system. Internationally, by promoting its own hardware, encryption, and software systems—often with financial incentives—China will have increasing influence over significant parts of the international system. For example, China started its own post-quantum cryptography algorithm standardization after NIST began its process. Limited to domestic submissions, China ultimately selected as winners one algorithm that had been submitted by a Chinese group to the NIST process but did not advance and two algorithms that are modifications of NIST submissions by an international consortium of researchers. China's future success in this regard will probably be a function of geopolitical factors rather than purely technological or economic ones—but it is possible (if not probable) that large portions of neighboring Asia as well as Africa and Latin America might well use Chinese communications systems in addition to or in lieu of U.S. or European ones.

China's emphasis on the development of its own technology and systems is principally the result of the country's political and strategic goals, and to a lesser extent of the reluctance of Western nations to rely on Chinese technology generally on the basis of security concerns. For example, the recent rejection by the United States and some of its allies of Huawei's 5G telecom equipment may well be a harbinger of the type of global divisions that will arise over Chinese technology. Similarly, recent efforts to strengthen integrated circuit production in the United States, although not aimed exclusively at China, are illustrative of efforts intended to preclude eventual Chinese dominance in a critical technology and could lead to China and the United States utilizing different electronic components.

Mistrust of Governments

Negative aspects of the Internet—the presence of disinformation, the absence of uniformly recognized curated news sources, and the resulting de-legitimization of institutions—all reflect and lead to increasing distrust of authorities, whether democratic or otherwise. In particular, Western democracies have themselves been grappling with a rise in the mistrust of governments (including in the area of surveillance) and a corresponding growth in the use of encrypted communications (both to avoid government surveillance and in response to general privacy and security concerns).

In large part as a result of social media, many nations, whether democratic or not, have been coping with an apparent rise in domestic extremism, with a seeming prevalence of conspiracy theories. In this regard, the availability of encryption affords extremist citizens' groups the ability to organize quickly and effectively, posing a threat to domestic order. A few agencies within the Intelligence Community that have a law enforcement function do have specific responsibility for collecting certain types of domestic intelligence within the United States; in addition, any ensuing increase in civil unrest among our allies and even adversaries might complicate or increase intelligence collection demands.

In the case of more authoritarian regimes, the potential for more encrypted communications is likely to lead a government mistrustful of its own citizens to ban or curtail effective encryption. Even in the case of democratic nations, the rise in the average citizen's use of encrypted communications increasingly has an effect on government's access, subject to applicable legal process ("exceptional access"), to the content of (and, in some cases, metadata about) communications. In general, such access could be expanded either because authoritarian regimes seek greater surveillance over their citizens, or because a more trusting or complacent citizenry is willing to accept such access as a price of greater perceived security and safety.

It is not clear that there is even a unified approach to the issue of exceptional access among the Five Eyes countries; at least some of the non-U.S. members have either adopted or are contemplating approaches to exceptional access which mandate that communications and device providers make available non-encrypted content upon relevant legal process. Indeed, it is possible, if not likely, that an increasingly fragmented outcome on this point could result. The implications for intelligence collection range from complexity and demands for greater resources to cope with more variegated targets, to possible ease of intelligence collection to the extent exceptional access schemes can be exploited. (See Box 4.3 for a further discussion of exceptional access and metadata.)

Economic and Security Concerns

Nations worry about not only content control but also perceived invidious economic advantages relative to other nations and their own national security risks arising from technology. The COVID-19 pandemic and recent cyber hacks from foreign sources have produced an increasing recognition in the United States that the country's well-being is more profoundly than ever dependent on the actions of other nations. There are now significant and growing policy debates about the extent to which nations and their allies should seek their own rules nationally—in opposition or at least without regard to those of other nations—to govern hardware, software, and data.

Economic competition and protectionism are likely to be factors promoting nationalistic solutions to communications technology, including encryption. These solutions could be seen especially as the United States and China seek economic hegemony, but also as the European Union seeks through antitrust and privacy regulation to limit U.S. technological dominance.

Another product of technology, cryptocurrency, presents additional challenges to governments. In particular, governments' inability (for technical or political reasons) to regulate cryptocurrencies could promote illegal activity generally, undermine taxing efficiency and support fragmenting forces throughout the global economy.

As to the national security concern, the U.S. reaction to Huawei's central involvement in 5G rollouts throughout the world—raising fears of Chinese government surveillance and collection of data—is a recent and highly visible example of how nations might address "supply chain risk" presented by hardware and software. That surveillance risk, augmented by general cybersecurity concerns arising from foreign cyber malefactors (as seen in the Solar Winds and Colonial Pipeline hacks of late 2020 and early 2021), are likely to solidify nationalist approaches in the United States and Western nations to cybersecurity risk management and encryption.

Apart from the effects of diverging developments in technology and the international standards that guide those developments, traditional economic and geopolitical forces might inevitably lead to a more fragmented world. Just how potent, and possibly irreversible, these forces of fragmentation might be is highlighted by the speed and scope of Russia's expulsion from much of the Western economy following the Ukrainian invasion and the extent of Russia's significant internal disengagement from Western news and social media. At the time of this report, it may be too soon to make a definitive judgment, but the profound international reaction to the Ukraine crisis may be a harbinger of future fragmentation. More broadly, traditional economic and geopolitical forces, coupled with the rise of China, could lead to the realignment of the 70-year-old postwar international system, and the concurrent diminution (for both domestic and international reasons) of U.S. leadership. The effect of this on encryption could be diffuse but pervasive. For example, political changes within the Five Eyes, as was the case with New Zealand nuclear policy in the 1980s, might weaken the alliance. Differing views on privacy and intrusion might also create future friction. The growth of regional blocs that join in taking different approaches to trade and regulation might also increase fragmentation.

Such a realignment could yield an international situation governed less by largely unchanging common interests than by ad hoc, transactional considerations—which in turn could make planning for the Intelligence Community more problematic. This could present novel challenges—in management, resource availability and allocation, and recruiting, among other things—for the Intelligence Community, which since its inception at the end of World War II has been focused for decades on a single critical problem set during the Cold War and, for the past two decades, on counterterrorism. Moreover, a significant component of the Intelligence Community's mission in the future will entail open-source information, requiring significantly greater involvement of the private sector that both holds much of that information and has the expertise to analyze it.

The increasing calls, notably from Europe and India, and more recently China, for nations to have protective ownership interest in the data of their citizens and businesses based on national security, privacy concerns and other grounds—so-called data sovereignty—are likely to have an effect on encryption and lead to more local control. Especially in Europe, notions about privacy and marketplace competition that differ from U.S. views are behind significant proposed restrictions on cross-border data flow and the regulation of online services.

It is easy to see how this could produce different approaches to encryption as well as data transmission and storage—but the effects on the Intelligence Community are mixed. For example, a requirement banning cross-border transfer of certain data and requiring in-country storage and processing might produce fewer opportunities for surveillance by interception in transit; on the other hand, it might facilitate surveillance because of the consolidation of target information in potentially accessible cloud or server storage controlled in the relevant nation. That might be offset, however, by requirements for local data storage that are often coupled with elevated cybersecurity standards, which could frustrate U.S. Intelligence Community access.

Cryptocurrencies potentially may create large disruptions in what information is available to governments, including intelligence agencies. The section above on Scientific Advances discusses technological developments related to cryptocurrencies and provides a brief discussion of policy-related issues. A full analysis of the possible regulations concerning cryptocurrencies is beyond the scope of the current study, but this report will briefly discuss possible impacts from low or high levels of such regulation.

If there is a low level of government regulation of cryptocurrencies, then many payments will be difficult or impossible to track by intelligence agencies or government agencies generally. Scientific advances might disable a particular cryptocurrency, but as long as cryptography works at all there will be ways to use it for payments that may be difficult or impossible to track. In the absence of financial records accessible to government, cryptocurrencies can be used to enable ransomware, money laundering, terrorist financing, financial sanctions evasion, and other criminal activities. Historically, regulatory regimes such as anti-money laundering efforts have tended to expand their coverage over time, as criminals have used unregulated cash-like instruments in place of regulated instruments such as cash or checking accounts.[33] Cryptocurrencies may also be used to place income and assets outside of the tax reporting system, potentially substantially reducing compliance with national taxation requirements. To the extent that cryptocurrencies evade such regulatory efforts, then intelligence agencies will have reduced access to financial information and finance-related illegal activity may increase significantly.

In the low-regulation scenario, there are market and technological factors that may lead to varying levels of growth in cryptocurrencies. In financial markets, enthusiasm about growth in cryptocurrencies has spurred rapid growth in investments in the sector, with total market capitalization above $2 trillion in early 2022.[34] In contrast to this scenario of rapid and sustained growth, government officials and others have expressed the view that cryptocurrencies are Ponzi schemes,[35] where early investors get high returns as total investment grows, but later investors lose their funds as the number of new investors diminishes. If the risk of such losses becomes large enough, then investment in cryptocurrencies may flatten or even decline over time. Along with these market unknowns, there are technological unknowns. As the sector grows, so does the incentive for attackers to steal or otherwise attack existing cryptocurrencies. Such attacks may be on the algorithms at heart of cryptocurrencies; in addition, as with cryptosystems generally, attacks may be possible even if the algorithms themselves remain resistant to attacks. In

---

[33] For example, casino chips were transported across national borders and cashed in a different jurisdiction, and so money laundering laws expanded to cover casino chips. P. Swire, 1999, "Financial Privacy and the Theory of High-Tech Government Surveillance," 77 *Washington University Law Quarterly* 461.

[34] C. Morris, 2022 "Crypto Market Cap Is Once Again Above $2 Trillion," *Fortune*, March 2, https://fortune.com/2022/03/02/crypto-market-cap-2-trillion.

[35] M. Singh, 2022, "Cryptocurrency Is Akin to 'Ponzi Scheme' and Banning It Is 'Perhaps the Most Advisable Choice,' Says India's Central Bank," TechCrunch+, February 15, https://techcrunch.com/2022/02/15/india-central-bank-cryptocurrency-ponzi-banning.

many settings there are weaknesses in the implementation of cryptosystems, so that attacks can succeed even if the encryption algorithm remains strong. The technological risk of successful attacks means there is uncertainty about future growth in cryptosystems, similar to the market risks if fraud grows too high or new investors shun the sector.

High government regulation may limit the growth and use of cryptocurrencies, in addition to possible limits on growth owing to market and technological factors. Governments have at least two regulatory strategies available to limit growth of cryptocurrencies outside of the traditional banking system. First, governments may make it expensive or illegal to convert from a cryptocurrency to the national currency, such as by prohibiting payments in cryptocurrencies to banks and other significant economic actors. Second, governments may issue and support (such as by regulation) a government-issued cryptocurrency. Under either approach, a larger fraction of transactions would remain visible to intelligence and other government agencies, and thus available for investigations into money laundering and other criminal activity. Early examples of government limits on cryptocurrencies include China's 2021 ban on Bitcoin and Japan's prohibition on use of anonymous cryptocurrencies.

This high level of government regulation could help combat crime but may also be accompanied by expanded government surveillance generally of financial activity. The multi-decade shift toward electronic financial transactions has already allowed companies and governments to track individuals and transactions at fine granularity. This trend toward financial surveillance may expand considerably with government-issued cryptocurrencies, or prohibition on use of privacy-protecting cryptocurrencies in commerce. High levels of financial surveillance would be contrary to the trend in many countries to provide expanded protection of personal privacy, against both corporate and government access to personal information. Defenders of financial privacy can point to historical abuses, including mandatory reporting in Nazi Germany of Jews' financial assets, as a step toward eventual seizure of such assets.[36]

> **FINDING 4.3:** It is difficult to predict what mix will occur of low or high levels of government regulation of cryptocurrencies. Low levels of regulation will be subject to criticism for facilitating criminal activity. High levels of regulation will be subject to criticism for excessive surveillance. Market and technological factors further make it difficult to predict future growth in the sector. Of this uncertainty, it is also uncertain the extent to which intelligence agencies will retain, increase, or decrease their access to financial, transactional information.

Climate Change

Last, on a more diffuse and speculative note, the effects of climate change, such as rising seawater levels and shifts to alternative energy production, might lead to a proclivity for local responses and adaptation. This could include possible reluctance by some countries to fund efforts to mitigate the effects in other countries. Moreover, countries might seek to address the resulting challenges in a national way given that the problem manifests itself in a territorial manner and the fact that effective global solutions are not immediately apparent or within reach.

> **FINDING 4.4:** Forces for both globalization and fragmentation will be present. Even if the committee were in a position to predict whether globalization or fragmentation were more likely to prevail, these trends are complex and interrelated. Some trends reinforce themselves and others prompt opposite reactions. Thus, it is difficult to determine which forces are likely to prevail on any given issue. In theory, this means that the Intelligence Community will need to be prepared for alternative extremes—for example, a world in which authoritarian governments weaken or ban encryption in ordinary communications, and a world in which governments support pervasive use of encryption citing privacy and security concerns. Because that preparation is impossible to sustain over any meaningful period, there will be a premium on accurate detection of trends at the earliest possible stage and managing the risk of an incorrect assessment.

---

[36] P. Swire, 1999, "Financial Privacy and the Theory of High-Tech Government Surveillance," 77 *Washington University Law Quarterly* 461.

FINDING 4.5: The Internet and increasing technological interdependence promotes globalization. The shared experience of individuals around the globe owing to information and communications being instantly and ubiquitously available is a powerful cause of international commonality. That factor, along with convergence of technologies, ever increasing global interdependence on all levels and across economic and political sectors, the continued growth of world trade and the expected increase in the role of the private sector, with its constant drive for efficiency and common standards, will tend to powerfully mold the world in a unified way, increasing the likelihood that nations around the world will take common approaches to issues relevant to encryption.

FINDING 4.6: Governmental regulation, for better or worse, of communications technology may lead to fragmentation on national lines. National security concerns have the effect, whether specifically intended or not, of creating competing national technologies—by limiting the exports of sensitive technology or by curtailing imports of equipment that may permit surreptitious surveillance by a foreign manufacturer or its government. Potent forces are present, for both beneficial and malicious reasons, that could predispose the global arrangement toward individual nationalistic or regional solutions to issues bearing on encryption. In many countries, there is growing support for "digital sovereignty," a term that can mean various things ranging from having regulatory decisions made nationally instead of by Silicon Valley, and support for protectionist trade policies, to segmenting the Internet by blocking communications with other countries. In addition, national regulations to promote online competition, enhance cybersecurity, curtail hate speech, and protect citizens' data privacy might well vary significantly around the globe and even in geopolitical regions where there might otherwise be commonality. A rise in citizens' mistrust of governments (especially in the area of surveillance) might lead to a corresponding growth in the use of encrypted communications (both to avoid government surveillance and in response to general privacy concerns). Moreover, individual countries or blocs of like-minded countries might impose (or continue to impose) substantive communications content requirements enabled by technological distinctions at national levels, including, for example, banning or discouraging end-to-end encryption (so as to permit government surveillance), or mandating a variety of governmental access to otherwise encrypted communications (perhaps through required turnover of encryption keys to authorities or insisting on the use of specified encryption schemes).

FINDING 4.7: In most cases, a common set of security protocols and cryptographic algorithms are used globally, and systems and networks today are largely interoperable. This may not remain the case; the factors that led to this interoperability are weakening, and pressures to create national and regional differences are growing.

Society and Governance Driver Summarized

Box 4.1 presents the considerations for the future of online communication and the two endpoints of globalization and fragmentation.

As will be seen in Box 4.2, increased fragmentation could affect the U.S. Intelligence Community in multiple ways, determined by governmental and societal changes. Box 4.3 expands on a possible future in which drivers create a situation in which communications "go dark."[37]

SYSTEMS DRIVERS: CHAOTIC VERSUS MATURE

The systems driver addresses the issues associated with the development and operation of technology products and systems that embed encryption. The extremes of the systems driver are chaotic and mature. Figure 4.3 defines what is meant by the systems driver and how a future would look if it existed at either extreme.

---

[37] R. Knake, 2020, "2019: The Beginning of the End of the Open Internet Era," Council on Foreign Relations, January 6, https://www.cfr.org/blog/2019-beginning-end-open-internet-era.

BOX 4.1
Society and Governance Driver: Which Direction for Online Communications: Globalization or Fragmentation?

Internet users can easily assume the Internet is necessarily "global": U.S. users can readily send messages to recipients globally and access many websites no matter their physical location. But such global access is far from inevitable. Users in some other countries are already restricted in what they can access, and the future ability to send messages and access websites globally faces obstacles. A fragmented communications infrastructure may be inevitable because of political and social pressures.

The development of public-key encryption is an important driver of today's decentralized global communications. It ensures that anyone can send a message securely because only the receiver has the private key needed to read a message's plaintext.

Widespread adoption of international consensus standards for encryption, as discussed in the context of the Scientific Advances driver above, also facilitates global communications. Encryption standards used in numerous countries are the basis of secure communications across international boundaries.

Indication of a shift toward fragmentation in encryption practices is already seen in the actions of countries like China and Russia to support or require use of government-selected encryption standards. If users are limited to their national encryption standards, international communications may need to be conducted using only algorithms that are lawful in both the sending and receiving country, perhaps replacing strong cryptography that meets international standards with weak cryptography such as the 40-bit encryption that was mandated by U.S. export controls in the past century, so that communications are readable by national authorities. For instance, major businesses such as banks face legal risks if they use applications or communications protocols that are illegal in a country where they operate. Worldwide secure, encrypted communications risk being replaced by weakly protected international communications or fragmented by blocs or regions, thanks to the imposition of national controls.

Another potential area of fragmentation concerns the public key infrastructure (PKI) "trusted root certificates" used by end-user software and devices to authenticate services (e.g., web services contacted over a secure connection by a user's browser). These explicitly trusted certificates form the basis of the trust hierarchy that ultimately lets any service on the Internet prove their identity in an interoperable fashion. When the PKI system works properly, it gives the user confidence that a website the user contacts is who it says it is—the user's bank, for instance, rather than a fraudulent site. Once authenticated, the PKI system enables secure, encrypted communications between the user and the bank. To date, the web PKI has been essentially globalized, in the sense that the major global browser and device manufacturers determine which certificates to recognize as trusted roots, and they will warn the user when a connection is made to a site that lacks a certificate that has a cryptographically valid signature (or chain of signatures) terminating in a trusted root.

The risks to this globalized PKI system are exemplified by the efforts of Kazakhstan since 2015 to create its own root certificate and mandate its inclusion in every browser and device used within its borders.[a] Such a root certificate, if trusted, could enable redirection of secure connections to fraudulent websites. Even for encrypted HTTPS traffic, this government-issued root certificate could enable the Kazakh government to conduct "man-in-the-middle" attacks—that is, to intercept, decrypt, and re-encrypt any traffic passing through systems it controlled. To date, browser vendors have resisted Kazakhstan's effort, such as by preventing the browser from recognizing the government-issued certificate even if the user seeks to use it. Going forward, however, nations might legally require users to deploy a browser with effects such as those sought by Kazakhstan. Already, some private enterprises deploy their own root certificate security policies and custom root certificates to perform similar monitoring functions on their own networks (sometimes mandated by regulatory requirements). In any of these scenarios, the previously globalized system for recognizing authentic websites could give way to more fragmented practices determined by which certificates a user's browser would recognize.

---

[a] R.S. Raman, L. Evdokimov, E. Wurstrow, J.A. Halderman, and R. Ensafi, 2020, "Investigating Large Scale HTTPS Interception in Kazakhstan," pp. 125–132 in *Proceedings of the ACM Internet Measurement Conference*, Association for Computing Machinery, http://www.acm.org.

BOX 4.2
Society and Governance Driver: Fragmentation Through Data Localization

In a globalizing Internet, data increasingly flows across borders, enabling access from multiple locations. Diminished international flows may result as countries impose more "data localization" requirements for four overlapping reasons: cybersecurity/national security, data sovereignty, privacy, and economic protectionism.

Some local cybersecurity laws are designed to limit data flows to other countries. China's strict requirements that prohibit data export by critical infrastructure industries offer a prime example, but numerous countries have cybersecurity laws, many of which impose limits on exporting sensitive data. Similarly, the Committee on Foreign Investment in the United States (CFIUS) also scrutinizes corporate acquisitions for sensitive data transfer. In 2019, CFIUS blocked Chinese owners' control of the dating site Grindr on national security grounds.

Recent calls for "data sovereignty" in many regions, including Europe and India, on national security and other grounds are a growing trend that challenges the long-standing U.S. commitment to an "open, interoperable" Internet and "free flow of data." The meaning of "data sovereignty" may vary but is commonly designed to enhance control by local regulators, rather than leaving decision-making to digital platforms. The advent of cloud computing, with stored data far from the country of origin, has created challenges for law enforcement or intelligence agencies that seek to use local procedures to access user information for criminal and other investigations.

Privacy concerns for personal data also drive geographic fragmentation. The 2018 European Union General Data Protection Regulation (GDPR) prohibits personal data transfer between the European Union and other countries unless safeguards "essentially equivalent" to the strict EU protections exist. In the 2020 *Schrems II* decision, the Court of Justice for the European Union raised doubts about standard contractual clauses designed to enable data transfers. Such developments may produce additional restrictions on data transfers between Europe and other regions. Laws modeled on European privacy rules have spread to more than 100 countries. If rigorously enforced, these Europe-style privacy laws could further fragment global communications and commerce.

Local economic advantage or protectionism is a fourth trend producing fragmentation. "Techlash" against the largest digital platforms, many U.S. based, may be an unstated motivation for national laws limiting data transfer internationally as nations and regions seek "national champions" to rival incumbent platforms. For example, the European Union's GAIA-X project encourages growing local cloud providers, stating: "An open digital ecosystem is needed to enable European companies and business models to compete globally. This ecosystem should allow both the digital sovereignty of cloud services users and the scalability of European cloud providers." Digital sovereignty and protection of user data are thus combined through support for explicitly European cloud providers.

Such changing patterns of data collection and transmission will affect all aspects of U.S. participation in the Internet, including the Intelligence Community.

BOX 4.3
Going Dark, Exceptional Access, and Communications Metadata

Going Dark

In response to increasing concerns of law enforcement agencies about the growing use of encryption by end consumers, the U.S. government in the early 1990s advocated the "Clipper" chip, an encryption device to be installed in telephones to support encrypted voice services but with the encryption keys held (escrowed) by the government. The Clipper initiative was abandoned in the face of considerable opposition by industry and some civil liberties groups. As the use of mobile phones increased, and with the advent of email and other Internet-based communications platforms, many of which used encryption at some level, law enforcement officials continued to assert that encryption was an enduring obstacle to their identifying and prosecuting criminal activity.

Current political debates about law enforcement access "going dark" center around three different uses of encryption. The first is where encrypted information is stored on a device such as a smartphone or memory equipment like a USB drive. Typically, in the first situation, the data are encrypted using a symmetric cipher, and the secret key is stored in a manner protected by the user's passcode. For example, Apple phones use hardware protection to prevent an external entity from "brute-forcing" (automatically trying every possible combination) the user's passcode to learn the secret key.

The second situation arises where law enforcement or intelligence agencies seek access to communications accessible outside of the device. The issues relating to government access vary depending on whether the service has access to plaintext, or whether instead there is end-to-end encryption. Major email services today deploy encryption from the sender to a server, and encryption from a server to the recipient; the services, however, retain access to the plaintext at the server. In such instances, a government agency can seek access to the content from the service provider. In the United States, for instance, government can access content held by a U.S. service provider upon a judicial order that there is probable cause of a crime.

Complexities can exist in this technological scenario, such as when a non-U.S. government seeks access to content held by a service governed by U.S. laws. In that case, the non-U.S. government must go through the time-consuming mutual legal assistance process to secure access to the email. The rules covering mutual legal assistance are complicated and vary according to which countries are involved. Further complications arise when a service provider is subject to conflicting laws on this point, such as where a company subject to U.S. laws has a server located in a country with conflicting rules. Nonetheless, where the communication is in plaintext at the server, legal procedures often exist for a government to seek access to the contents of the communication.

The third situation is end-to-end encryption, which is commonly used today for messaging apps. In end-to-end encryption, two end users use an algorithm like Diffie-Hellman key exchange to establish shared secrets directly with each other and encrypt messages so that only the recipient can read them; the server routing these messages would have access only to the ciphertext. Even in some cases of end-to-end encryption, the user's expectations might not be met, and authorities might be able to gain access. For example, the user's messages on a cell phone might be saved in plaintext form and accessible to authorities who gained access to the phone even though the actual communication to the recipient was encrypted.

Law enforcement agencies have continued to advocate for "exceptional access" to encrypted information, for at least the three scenarios described above: encrypted storage on end-user devices, end-to-end encryption from message originator to recipient, and encryption in transit to a communications server in a different country, where the government often lacks the ability to gain access. Over the past 5 years, there have been proposals by private researchers for schemes that would enable such access in at least some scenarios of interest. The proposed schemes vary in the roles for and degrees of control by government and device vendors and the level of control over the exercise of access rights. Although the issue is highly contested, there is a recognition that there are differences between devices, where physical possession of the device may permit heightened safeguards, and encryption in transit.

Civil liberties groups, many encryption researchers, and most device vendors have opposed the introduction of exceptional access mechanisms, asserting that such access would violate privacy rights or introduce security weaknesses that could be exploited by others including hostile intelligence services. In addition, many opponents point out the existence of possible alternatives, saying that "metadata" (typically not encrypted) has the potential to provide law enforcement with a wealth of information about communications, albeit not the plaintext of encrypted communications.[a,b]

One additional issue concerns the efforts of governments to persuade or require services to avoid or minimize use of end-to-end encryption. For non-end-to-end encryption, governments can access plaintext through the service provider, in democratic nations typically with a court order. Governments have less reason to push for exceptional access if they can gain access through such court orders. An increasing area of policy contention, then, may be government efforts to favor non-end-to-end encryption instead of end-to-encryption.

continued

BOX 4.3 Continued

Metadata

Metadata is the name that has been given to information about information, as distinguished from the content of the information itself. For example, in an encrypted messaging application, the identities of sender and receiver, the length of the message, and the times of transmission and reception would be classified as metadata. This form of metadata is similar to the "messaging externals" that are the target of traffic analysis. Typically, some amount of metadata will inevitably be created by any application that communicates or stores encrypted information and that metadata is subject to seizure or interception by law enforcement even if the encryption cannot be defeated. Whether law enforcement access to metadata is a sufficient substitute for access to the plaintext underlying encrypted information is a topic of ongoing debate that this committee is not charged with resolving.

Additionally, there are cryptographic methods that may be employed to minimize the metadata revealed by network communications. The Tor networking protocol is one such example; Tor uses encryption to route data through a network in such a way that no entity should be able to link the source and destination of communications. However, the Tor protocol introduces a high overhead cost to the communication it is protecting. Improving the robustness and performance of anonymous communication networks is an area of active research and development.

Resolving the Debate

Encrypted and anonymous communication techniques have both positive and negative uses: they enable users to circumvent network censorship in countries such as China and Iran, but also have enabled crime, the "dark web," data exfiltration, and espionage. Intelligence agencies are beneficiaries of those techniques when they use them to protect raw intelligence and sources and methods, but they are threatened by their use when they facilitate actions by their adversaries.

The exceptional access debate in the United States is long-standing and there is no clear way to predict if or how it will be resolved. The United Kingdom and Australia have passed legislation that appears to require communications carriers and perhaps technology vendors to facilitate access to encrypted information, but the state of implementation is not clear. China like many other nations does not discuss exceptional access as such although Chinese laws and regulations do appear to grant the Chinese Ministry of State Security a significant level of access to information technology systems.[c]

Last, the focus of exceptional access in the United States has historically been around the law enforcement mission, with officials pointing out that state and local agencies do not have the resources to attempt to overcome encryption and that there is no substitute in criminal prosecutions for evidence that may be encrypted in communications. By contrast, the voice of the U.S. Intelligence Community in the encryption debates has been less intense, first because the Intelligence Community is legally precluded from focusing on U.S. citizens and second because the Intelligence Community does not share the responsibility for possessing evidence that typically must later be publicly released in a court proceeding. Even more significantly the Intelligence Community, with its greater resources, is often in a position to develop the necessary insight into targets' behavior and communications by multiple means. Nonetheless, the Intelligence Community, too, has struggled with powerfully encrypted communications. For this reason, some Intelligence Community officials have characterized the problem for them as "going dim" rather than "going dark."

---

[a] H. Abelson, R. Anderson, S.M. Bellovin, J. Benaloh, M. Blaze, W. Diffie, J. Gilmore, et al., 2015, "Keys Under Doormats: Mandating Insecurity by Requiring Government Access to All Data and Communications," MIT-CSAIL-TR-2015-026, July 6.

[b] Encryption Working Group, 2019, "Moving the Encryption Policy Debate Forward," Carnegie Endowment for International Peace, Washington, DC, September.

[c] INSIKT Group, 2019, "China's New Cybersecurity Measures Allow State Police to Remotely Access Company Systems," Recorded Future blog, February 8, https://www.recordedfuture.com/china-cybersecurity-measures.

Driver Details: Systems

The manner in which encryption technologies are implemented

Below are example polar extreme characteristics of the Systems driver:

Chaotic		Mature
Security-critical systems continue to increase in complexity without commensurate reductions in defect densities.	⟷	Successful efforts to reduce complexity and defect densities of security-critical portions of systems.
Development practices prioritize low costs, feature-rich products. More secure products cannot compete or are a small part of the market.	⟷	Development practices for mainstream products prioritize security, resulting in higher costs and removal of nonessential risky features.
Developers create risks, but consequences largely borne by end users.	⟷	Developer investment in security is commensurate with downstream risk.
Relying parties continue to have poor visibility into risks; certifications are funded by vendor marketing and approve vulnerable products.	⟷	Customer gains practical ways to accurately gauge risks.
Continued widespread use of memory-unsafe programming and hardware architectures that do not prioritize security.	⟷	Transition to safer programmer languages; formal methods help secure cryptographic implementations and other security-critical code; transistor gains from Moore's Law enable on-chip security hardware.
Messy transition to quantum resistant public key algorithms. Competing standards, messy transitions.	⟷	Broad consensus around NIST approved algorithms, orderly adoption, predictable quantum computer developments.
U.S. educational system produces insufficient cryptography and data security talent, and it becomes more difficult for the U.S. to attract talent from other countries.	⟷	U.S. remains a major destination for global talent, and increases its ability to produce experts as well as basic knowledge among engineers and end users.

FIGURE 4.3 Visual representation of space defined by Systems driver.

Most attacks that compromise cryptographic systems exploit implementation weaknesses. This is partly owing to the strengths of cryptography; widely used algorithms are typically based on well-researched mathematical problems, designed with large safety margins, and carefully reviewed before gaining trust. In contrast, the security-critical elements of a typical computer system are so large and complex as to virtually ensure the presence of errors that lead to exploitable vulnerabilities.

The balance between systems-related risks and algorithm-related risks has shifted over the past decades. Moore's law and related advances have enabled a dramatic decrease in the cost of computation and a corresponding increase in the complexity of computation systems. Historically, cryptographic algorithms were severely constrained by the limited computational complexity that was practical during encryption and decryption, and as a result were often breakable. In the 1980s, constraints on computation power started to ease, but export control regulations, the 56-bit key size of the Data Encryption Standard (DES)[38] and general unavailability of cryptographic knowledge limited the strength of most commercial implementations.

Today's systems fail as a result of exactly the opposite problem; the underlying algorithms are extremely robust, but complexity in the systems that use them results in vulnerabilities that adversaries routinely exploit to break or bypass the cryptography. The real-world security properties of systems are typically determined by the weakest link—and there is a lot of room for mistakes. Typical computing devices today comprise billions of transistors and millions of lines of software distributed across operating systems, firmware, device drivers, software

---

[38] The first encryption algorithm standardized by NIST for use by the private sector and the civil agencies of the U.S. government was called the Data Encryption Standard, commonly referred to as DES.

libraries, and application code. Trends for greater network connectivity, more sharing of resources between tasks, and exposure of more sophisticated programming interfaces enable cost savings and powerful features, but also make vulnerabilities more likely and more exploitable. The complexity and defect densities for typical computing systems today are sufficiently high that vulnerabilities are essentially certain.[39] Unless there are reductions of many orders of magnitude in complexity and/or defect densities, vulnerabilities will remain widespread.

Security vulnerabilities are different from other kinds of defects. Normal bugs result in observable misbehavior that motivates remediation efforts. For example, if an important system fails to boot or crashes, people will notice and fix the problem. By contrast, security flaws tend to be invisible to developers, to testers focused on the functioning of system features, and to end users. Excess complexity makes vulnerabilities even harder to find; even vulnerabilities that seem glaringly obvious once highlighted often lie undiscovered for years in neglected parts of a codebase that work well for non-malicious inputs. In addition, fixes for non-security bugs are individually beneficial, and provide benefits that scale with the number of bugs fixed. In contrast, security efforts are more like trying to make a sieve watertight. If a product has thousands of security bugs, a security effort that fixes 75 percent would appear successful in terms of the number of vulnerabilities, yet might make no practical difference if attackers still have plenty of ways in. Security bug responses can also be chaotic, and therefore more prone to introducing new problems, as developers must rush to implement fixes before vulnerabilities are exploited (or to stop an active attack). Information technology (IT) departments receive updates from numerous vendors and face a constant, and often overwhelming, deluge of urgent fixes. Even planned updates to introduce new security measures can impose costs on end users, such as usability challenges associated with extra authentication steps.

Of course, defenders also strive to reduce risk. These efforts include fixes to specific security bugs, broader mitigations for specific categories of attack, and monitoring of operational systems to help detect or block intrusions. This results in an ongoing "tit-for-tat" dynamic, where offensive efforts try to evade defenses, and defensive efforts try to mitigate attacks. While some kinds of defensive responses can occur quickly, others can take many years because new research is needed, hardware has to be replaced, or the vulnerability arises from upstream components in complex supply chains. At a higher level, perceptions of risk and externally imposed requirements determine each organization's willingness to accept costs, reduced functionality, inconvenience, and other trade-offs in pursuit of security initiatives. In theory, market forces might create some semblance of a balance between offense and defense over longer periods of time, but the level of investment will likely reflect the cost/benefit trade-offs for development organizations—who have traditionally been mostly isolated from the damage that breaches cause to downstream organizations and individuals. Furthermore, unlike risks such as credit card fraud where rates can be monitored and managed over time, attackers exploiting systems vulnerabilities often combine stealth or surprise with speed to make enormous gains before defenders can react. Such issues are especially acute in IoT systems where cyber-physical attacks have the potential to damage real-world equipment such as components of electrical grids or transportation systems.

Chaotic Versus Mature Systems

The Systems driver concerns the security properties of systems using cryptography. These properties include the security logic itself, such as algorithms, protocols, hardware, and software. A great deal of software whose purpose is unrelated to security is also included because it may harbor bugs (such as buffer overflow vulnerabilities) that attackers can use to bypass or break the cryptography or other security mechanisms. The Systems driver also considers the development and operation of systems, including practices and tools such as software engineering, testing, key management infrastructure, operator knowledge and training, monitoring systems, code update/remediation capabilities, and acquisition processes.

The "chaotic" extreme represents hardware/software systems and development methodologies that are likely

---

[39] O.H. Alhazmi, Y.K. Malaiya, and I. Ray, 2006, "Measuring, Analyzing and Predicting Security Vulnerabilities in Software Systems," *Computers and Security*, https://doi.org/10.1016/j.cose.2006.10.002.

to result in exploitable security vulnerabilities. A chaotic scenario may involve development teams making liberal use of third-party hardware and software components without thorough security diligence, programmers working without security experience or guidance, rapid addition of new functionality requested by customers without careful consideration of security implications, and high complexity in security-critical components. The operators and administrators of chaotic systems may be inattentive to security practices or overwhelmed by the task of managing the systems securely. Vendors producing chaotic systems often have poor awareness of risk and may advertise false or misleading security claims out of ignorance or self-interest. Ignoring security, chaotic approaches can offer large benefits for customers, including low costs, rich feature sets, and rapid upgrades.

The "mature" case assumes development processes designed to minimize the likelihood and impact of vulnerabilities. The number of security defects in a system scales with both the complexity of the system (e.g., lines of code) and the defect density (e.g., defects per line of code). As a result, "mature" scenarios require combining technical and human measures to reduce defect densities with a commitment to controlling the complexity of security-critical components. These reductions will invariably involve significant trade-offs, such as omitting desirable features (or disabling them by default), accepting performance overhead from technologies that isolate sensitive computations and data, and spending a lot more time and money on development and testing.

With some notable exceptions, today's systems are chaotic. Mainstream CPUs, operating systems, and major applications are so complex that even today's more careful development teams have no realistic chance of producing vulnerability-free designs. The effort required to find vulnerabilities is often within the capabilities of a determined and skillful individual, and probably always within the capabilities of a sophisticated nation state or other advanced persistent threat.[40]

Factors Leading to Chaotic or Mature Systems

The prevalence of chaotic systems is not owing to malice; technology companies do invest in security and would like to do better. However, the dominant factors driving technology development today strongly favor chaotic systems. Market share and profits are driven by characteristics of chaotic systems, including fast time to market, rich feature sets, and low development costs. The costs of insecurity and the operational burden of security measures that do get adopted largely fall on users rather than developers. Developers' economic incentives to sacrifice time-to-market and product features for security are comparatively weak, such as the desire to avoid bad publicity and customer complaints. Chaotic vendor-operated services (such as cloud applications) face similar pressures, although customers have a somewhat different mix of burdens and have less visibility into server-side vulnerabilities and operational risks.

Information asymmetries between developers and relying parties also favor chaotic systems. The security risks of technology products are often opaque to end users, so customers cannot easily compare products. Developers have strong commercial incentives to claim trustworthiness, and rarely face penalties for touting the security of vulnerable products. Customers' pricing expectations reflect the cost required to implement new functionality, not the (potentially far greater) costs of mitigating the risks created by the added complexity. If markets for technology products or government regulations do not place sufficient value on mature systems (or if customers cannot tell the difference or are unable to switch), vendors who produce them will fail or be relegated to high-priced niche offerings.

Technical advances that make mature systems easier to develop could result in significantly different outcomes. In addition, economic, political, and regulatory factors (as might be imposed after a humanitarian disaster caused by a computer security failure or if the number of small disasters becomes large enough that a major government

---

[40] The Defense Science Board report on Resilient Military Systems provides a useful perspective on adversary capabilities. See Department of Defense, 2013, *Task Force Report: Resilient Military Systems and the Advanced Cyber Threat*, Defense Science Board, Office of the Under Secretary of Defense for Acquisition, Technology and Logistics, Washington, DC, January, https://dsb.cto.mil/reports/2010s/ResilientMilitarySystemsCyberThreat.pdf.

decides to take dramatic action) could also change the equilibrium; such factors might vary by country because they would likely be defined by national interest, including legislation.

Improving System Security

The following sections discuss potential drivers that could change the status quo and make systems more mature.

Incentives, Transparency, and Development Process

Developer "Skin in the Game" The direct costs of security breaches today overwhelmingly fall on end users. Although security outcomes can affect technology developers' reputations, the effects are small enough that security initiatives tend to be very limited in their ability to affect development time, costs, performance, or features. Security investments would likely increase if legal or regulatory changes shifted liability to vendors—but such a shift could also bring enormous downsides, depending on nuances such as how liability is allocated. For example, the consequences of a security vulnerability depend on numerous contributing factors, such as whether (or which) adversaries discover the issue, how quickly it is fixed, end-user actions outside the developer's control, and so on. Liability obligations would particularly affect low-cost or free/open source software developers as they do not have a revenue source to fund either security efforts or the liability. Still, some countries may impose liability obligations anyway.

Metrics, Transparency, and Disclosure End users generally cannot accurately assess the security properties of technology products. Even when the source code and design documents are available, it takes substantial effort and expertise to perform a meaningful security evaluation. Optional or mandatory disclosure rules for vendors might help, but it is difficult to design rules that provide meaningful insights and encourage improvements, rather than devolving into paperwork exercises. If disclosures matter (i.e., affect customer behavior), then vendors will try to maximize their competitive advantages, and some will prioritize security resources on efforts that boost their scores over better ways to improve security. Some will be even more aggressive in seeking competitive advantages—for example, undermining competitors' scores, litigating if they do not like their score, and lobbying for changes in the system itself.

Third-party certifications also face difficult conflicts of interest; evaluations are typically funded by vendors, so evaluation schemes and testing labs survive by minimizing evaluation costs and approving products. If a metric gains enough traction to affect customer behavior, vendors will invariably seek competitive advantages by using the same sorts of tactics that they would use in response to traditional metrics such as performance benchmarks. Still, imperfect approaches may prove beneficial. The Biden administration's 2021 Executive Order 14028 seeks to make a start by requiring that vendors to the U.S. government implement a relatively specific set of practices during development and be prepared to show artifacts demonstrating their implementation to government customers. Although government efforts to mandate product security have historically met with limited success,[41] industry response to the Executive Order has been relatively positive and the Executive Order will likely cause vendors to the government to adopt these practices.[42]

---

[41] S.B. Lipner, 2015, "The Birth and Death of the Orange Book," *IEEE Annals of the History of Computing*, April–June.

[42] See the agenda, summary, and position papers for the NIST workshop in response to the Executive Order at NIST, 2021, "Workshop and Call for Position Papers on Standards and Guidelines to Enhance Software Supply Chain Security," updated June 11, https://www.nist.gov/itl/executive-order-improving-nations-cybersecurity/workshop-and-call-position-papers.

Secure Development Processes

Many software vendors have sought to improve the security of their products by instituting development processes that specify security-related requirements for development teams.[43] Those range from threat modeling to discover design-level vulnerabilities[44] to specific requirements, such as tools, coding constructs, and libraries. These processes typically combine human activities (such as code reviews) and automated tools (such as static analysis and fuzzing tools, as discussed below) to detect potential vulnerabilities and to assess whether delivered code meets the process requirements.

Secure development processes aim to help organizations pursue continuous improvement: the root causes of reported vulnerabilities are investigated and, where feasible, process requirements, tools, and developer training are updated to prevent the recurrence of similar vulnerabilities. Over time, the application of secure development processes has enabled organizations to reduce the frequency and severity of reported vulnerabilities.[45]

There are challenges to the adoption of secure development processes. Organizations with limited resources may not be able to invest in creating a process or acquiring and tailoring the necessary tools. In other cases, projects may not prioritize security or lack the centralized control necessary for consistent adoption of a secure development process. However, adoption of such processes continues to grow in response to customer pressure and U.S. government requirements (see discussion of Executive Order 14028 below).

Technologies for Improved Security

Advances in Formal Methods Formal methods is the name given to the application of mathematical techniques to prove the correctness and other properties of software. Although proven-correct software requires specialized skills to develop, and far more labor than conventionally developed software, some cryptographic algorithm and protocol implementations are relatively well suited to formal verification. If current progress continues, formally verified high-performance open-source libraries fully implementing protocols such as SSL/TLS (including the cryptographic algorithms, protocol handling, certificate processing, etc.) could be developed.

However, formal verification methods are not currently able to scale to the complexity of today's typical computing systems. A good example is Sel4—a secure operating system kernel that has been formally verified.[46] Although an important effort, Sel4 has less than 0.1 percent as many lines of code as the Linux kernel and lacks much of the functionality found in a typical operating system.

Vulnerabilities can hide in gaps between proofs and reality. Still, a breakthrough that allowed the economical application of formal methods to larger software systems could significantly change the security of the products that people and organizations use. Even incremental progress in the field might lead to improvements that would make attackers' search for vulnerabilities more difficult.

---

[43] NIST, 2022, "Secure Software Development Framework," Computer Security Resource Center, Information Technology Laboratory, updated February 3, https://csrc.nist.gov/Projects/ssdf.

[44] A. Shostack, 2014, *Threat Modeling: Designing for Security*, John Wiley & Sons, Indianapolis, IN.

[45] M. Miller, 2019, "Trends, Challenges, and Strategic Shifts in the Software Vulnerability Mitigation Landscape," presentation at BlueHatIL, February 7, https://github.com/microsoft/MSRC-Security-Research/tree/master/presentations/2019_02_BlueHatIL, accessed October 15, 2021.

[46] G. Heiser, 2020, "The SEL4 Microkernel: An Introduction," Revision 1.2 of 2020-06-10, The sel4 Foundation, https://sel4.systems/About/seL4-whitepaper.pdf.

Memory-Safe Programming Languages and Development Tools Memory safety bugs cause the majority of vulnerabilities in many large software projects (including Microsoft's product lines,[47] Android,[48] Chromium,[49,50] iOS, MacOS,[51] Ubuntu Linux,[52] and Firefox[53]). Operating systems, device drivers, software libraries, and other critical software are largely written in programming languages that lack memory safety, and memory safety programming errors continue to cause a very significant fraction of product security vulnerabilities. For example, a single missing bounds check on an array access in the C programming language can allow attackers to compromise an entire process (including logic and data never legitimately accessed by the buggy function).[54] More modern programming languages have the potential to reduce dramatically the harm of these and many similar software bugs. Today, many software teams are transitioning to the Rust programming language, which can guarantee memory safety at compile time.

Static Analysis and Fuzzing Static analysis tools identify potentially vulnerable code patterns in software, while fuzzing tools repeatedly execute software with varying inputs to detect memory safety violations or other misbehavior. These tools are used by many software development teams and have identified many vulnerabilities in commercial and open source products. The widely used tools identify suspicious code patterns but can miss some vulnerabilities. In contrast, "sound" static analysis tools are guaranteed to catch all vulnerabilities within the categories addressed by the tool, but for most programming languages they are not widely used today as they tend to find large numbers of false positives that are time consuming for developers to address. Executive Order 14028, mentioned above, effectively requires that developers use static analysis tools and fuzzing as they build software that will be sold to the U.S. government. Improvements in static analysis and fuzzing tools are important, and sound tools in particular could have a large impact if they can be designed with acceptable false positive rates.

Hardware Security Advances Hardware security plays a major role in many applications of cryptography. For example, tamper-resistant hardware can help protect cryptographic secrets from adversaries who steal or gain access to physical devices. Within complex systems, cryptographic operations can be implemented in enclaves or other hardware that operates independently from more vulnerable elements such as the main CPU, memory, and operating system. In other cases, cryptographic accelerators or specialized CPU instructions may improve performance or power consumption but do not add any extra protection.

Cryptographic hardware can take many physical forms, ranging from tiny logic blocks on a larger semiconductor chip to large appliances. Some technologies also blur lines between hardware and software. For example, the security may appear to be isolated in hardware, but in practice depend on firmware, microcode, FPGA bitstreams, or other externally supplied configuration data. If done well, cryptographic capabilities implemented in dedicated silicon, where security-critical logic is not shared with untrusted operations, can provide the highest levels of security. At the same time, backdoors in hardware are extremely difficult to detect.

---

[47] MSRC Team, 2019, "A Proactive Approach to More Secure Code," Microsoft Security Response Center, July 31, https://msrc-blog.microsoft.com/2019/07/16/a-proactive-approach-to-more-secure-code.

[48] J. Vander Stoep and C. Zhang, 2019, "Queue the Hardening Enhancements," Google Security Blog, May 9, https://security.googleblog.com/2019/05/queue-hardening-enhancements.html.

[49] K. Serebryany, 2018, "Hardware Memory Tagging to Make C/C++ Memory Safe(r)," Github, December, https://github.com/google/sanitizers/blob/master/hwaddress-sanitizer/MTE-iSecCon-2018.pdf.

[50] The Chromium Projects, "Memory Safety," https://www.chromium.org/Home/chromium-security/memory-safety, accessed October 15, 2021.

[51] P. Kehrer, 2019, "Memory Unsafety in Apple's Operating Systems," July 23, https://langui.sh/2019/07/23/apple-memory-safety.

[52] Ubuntu Security, 2020, "USN-4360-1: json-c Vulnerability," May 14, https://ubuntu.com/security/notices/USN-4360-1.

[53] D. Hosfelt, 2019, "Implications of Rewriting a Browser Component in Rust," Hacks.Mozilla (blog), Mozilla, February 28, https://hacks.mozilla.org/2019/02/rewriting-a-browser-component-in-rust.

[54] MSRC Team, 2019, "A Proactive Approach to More Secure Code," Microsoft Security Response Center, July 31, https://msrc-blog.microsoft.com/2019/07/16/a-proactive-approach-to-more-secure-code.

Improvements in semiconductor technologies have dramatically reduced the costs of adding security-focused logic into chips. As a result, large chips can combine many processors, some of which may be dedicated to security functions. Discrete devices for tasks such as network encryption, key management, user authentication, and security monitoring would fill a large rack and cost thousands of dollars. Implementing the same logic as isolated on-chip logic blocks adds negligible manufacturing cost—in addition to providing superior power consumption, convenience, and tighter integration.

Semiconductor advances are the most beneficial in areas where security can be improved by adding logic to devices. For example, adding memory encryption can reduce the need to trust external DRAM, and CPU features like CHERI[55] can reduce the exploitability of programming errors. Enclaves and related software isolation mechanisms are also an area of active interest, although designs that share large amounts of circuitry between protected and normal modes have struggled with vulnerabilities arising from the large attack surface area. As transistor prices fall, the cost incentive to reuse circuitry will decrease, while allowing chip vendors to add many security blocks optimized for different use cases and customers. The costs of logic implemented in hardware and software will also become more similar, where the up-front engineering costs dominate while the incremental costs of manufacturing and distributing the logic become negligible.

Application-Specific Solutions Even if the bulk of the market is chaotic, there may be islands of maturity. Security components such as secure enclaves on endpoint devices, cryptographic credentials such as smart cards or tokens, and robust encryption libraries may make it possible for organizations to create assured solutions for specific tasks. Similarly, governments and companies that are at high risk and have large security budgets can also produce or buy custom high-assurance solutions, although these solutions will typically be much more expensive and less attractive because they are much slower to adopt new functionality than mass-market alternatives. Larger countries will continue to invest in customized solutions—for example, to protect sensitive communications, classified data, intelligence operations, and weapons systems.

The success of these application-specific solutions depends on many factors beyond security, including the complexity of the integration and the incentives for adoption. For example, a technology that requires reworking a system's user interfaces, databases, and so on, is much more challenging than one that connects easily into established hardware or software interfaces. Some of these solutions can be built using relatively straightforward cryptography but could have large implications for the Intelligence Community. For example, a secure remote access system that facilitated a transition to distributed workforce would have major implications for hiring, facilities, culture, and employee oversight.

Integration and Operations

System Architecture Enterprise information technology systems typically encompass a variety of client and server computers along with networks that interconnect enterprise systems with each other and with outside systems connected via the Internet or private networks. In many cases, enterprises also rely on servers in the "cloud" that provide computing and/or storage services.

The architecture of an enterprise IT system defines the components that make up the system and technical requirements for interconnections. Effective security needs to be central to the architecture. An effective security architecture can help to mitigate (though not eliminate) shortcomings in the security of the individual products that are the components of a system.

---

[55] R.N.M. Watson, J. Woodruff, P.G. Neumann, S.W. Moore, J. Anderson, D. Chisnall, and N. Dave, 2015, "CHERI: A Hybrid Capability-System Architecture for Scalable Software Compartmentalization" in *Proceedings of the 36th IEEE Symposium on Security and Privacy ("Oakland")*.

A fundamental principle of security architecture is "least privilege"—the idea that no user, computer program, or component should be granted more access than it requires to perform its function.[56] Some designers of modern IT systems have adapted least privilege to a concept called "zero trust," that aims to build enterprise IT systems that do not make any assumptions about the security of the network. Zero trust eliminates or diminishes reliance on a network firewall to create a trusted network. As a result, users are allowed to use untrusted networks (such as potentially compromised corporate networks or public Wi-Fi access points) to perform sensitive tasks. In this model, applications and servers must carefully ensure that all accesses are authorized.

Implementations of zero trust architectures seek to ensure that every attempt to gain access to a resource (a file, a device, or a communications channel) must be authorized and the user or system attempting access must be strongly authenticated. All resources in transit over a network or in storage must be protected from unauthorized access. Encryption is critical to the implementation of zero trust architectures: authentication relies on encryption to implement digital signatures and data protection relies on encryption to provide secrecy and prevent unauthorized access.

Supply Chain Security Computer systems and networks bring together products and components created by a worldwide ecosystem of commercial vendors, open-source developers, manufacturers, and integrators. If systems do not take extraordinary efforts to isolate individual components, a vulnerability in an "outsourced" piece of hardware or software can have a devastating effect on the security of the entire system. Such a vulnerability can result from a malicious actor or inadvertent error at any stage in the supply chain or involve misunderstandings across components or layers of abstraction. The "SolarWinds"[57] intrusion disclosed in late 2020 is an example of a supply chain attack that was reported to have resulted from a malicious modification.

Governments and vendors have been seeking to establish policies and processes to prevent or defeat supply chain attacks. Executive Order 14028 mentioned above is a recent example. But supply chains are complicated, and the problem is a difficult one that plays a significant role in driving systems in a chaotic direction.

End-User Operational Security and Risk Management The inevitability of security failures does not mean that defense is pointless. Organizations can raise the bar against successful attacks by adopting mundane practices such as patching and updating software,[58] configuring their networks to require strong authentication of users and authorization of access, continuous backup and checkpointing, collecting and protecting logs of users' and programs' actions, and rapidly fixing vulnerabilities that are detected. At a minimum, such practices are effective against attackers with limited competence and resources (such as many ransomware perpetrators), while increasing the development cost and reducing the operational lifetime of attacks. Effective defenses against "low end" attackers can also help organizations reserve their best defensive resources and people for use against sophisticated adversaries.

Possible evolutionary paths for risk management include investing heavily in thoroughly verified systems to reduce the number of vulnerabilities, or reducing reliance on higher-risk components, deploying redundancy to combine less reliable systems into something better where failure modes are mostly independent (an approach that the National Security Agency has advocated in its "Commercial Solutions for Classified" program[59]), and where necessary combining many insecure components with insecure monitoring systems and unreliable people but managing to detect enough attacks to "muddle along."

Recovery and resilience can be important although they do not work effectively in cases where the secrecy of

---

[56] J.H. Saltzer and M.D. Schroeder, 1974, The protection of information in computer systems, *Communications of the ACM* 17(7).

[57] Center for Internet Security, "The SolarWinds Cyber-Attack: What You Need to Know," last updated March 15, 2021, https://www.cisecurity.org/solarwinds.

[58] Organizations' reliance on obsolete information technology products that have not been designed to stand up to new threats or benefited from secure development practices, or even are no longer provided with security patches by their developers, has proven to be a major cause of security incidents and a major concern for information security managers. See https://slate.com/technology/2018/06/why-the-military-cant-quit-windows-xp.html.

[59] National Security Agency, "Commercial Solutions for Classified Program (CSfC)," https://www.nsa.gov/csfc/, accessed October 15, 2021.

information is paramount (e.g., it is not possible to "recover" from a leak such as the Office of Personnel Management [OPM] breach[60]). Movement of computing to the "cloud" is a trend that will continue at least for the near future. Cloud services tend to be better managed and protected than many individual enterprise systems but the consequences of a successful attack on a cloud provider could be extremely serious.

FINDING 4.8: In every scenario, bugs in software and operational errors are the weakest links in security.

FINDING 4.9: Communications and storage depend on a software stack: hypervisor (a program that allows a computer to run several operating systems simultaneously), operating system, libraries, and application. While quantum computers or mathematical advances are important research topics, bugs or operational mistakes in this stack are the biggest source of system insecurity. Exploiting these errors is, and likely will remain, the biggest opportunity for offense, and minimizing them the highest priority for defense and risk management.

Education, Workforce, and Training

People with the right knowledge and expertise are essential for the reliable implementation and operation of cryptographic systems over the long term. Within U.S. universities, "security" is frequently considered a niche specialty or elective, much like graphics or computer games, rather than a cross-cutting issue of fundamental importance to all systems at the same level as algorithms or data structures. Although some self-taught individuals develop significant expertise in security, and some companies invest substantial sums in security education and training for their developers and operational staff, many do not. Some programmers have little enthusiasm for security, such as one survey respondent who commented, "I find the enterprise of security a soul-withering chore and a subject best left for the lawyers and process freaks. I am an application developer," while another said, "I find security an insufferably boring procedural hindrance."[61]

Thus, future U.S. scenarios are likely to lean toward "chaotic" without substantial improvements in our educational system and developers' full commitment to reliable, trustworthy implementations of hardware and software systems. While the Internet has enabled broader dissemination of knowledge about cryptography and security, the high-level research training and careers required to understand and contribute to the state of the art require real investment in research infrastructure from government and industry.

FINDING 4.10: The United States needs far more data security expertise than is currently available, and these needs are growing substantially. The failure to meet these needs could have significant and widespread ramifications both for national security and the private sector. All software developers and computer scientists require basic competence in computer security. In addition, a growing number of people will require deep expertise in security. The required skills are not easy to teach, as students need both security-focused knowledge and a deep technical knowledge across multiple subjects and layers of abstraction. If the U.S. educational system does not meet these needs, or if the United States becomes a less attractive destination for students, researchers, and entrepreneurs born in other countries, the shortage will be much worse. Technological changes may rapidly increase demand for rare skills or may reduce demand by enabling tasks that currently require exceptionally skilled individuals to be performed by a broader range of people

FINDING 4.11: Practical knowledge about the security of cryptographic systems will continue to be widely disseminated across the globe. Effective work (offensive or defensive) can be performed by a few skilled individuals. As a result, unlike areas where a country can obtain dominant capabilities by incurring costs that

---

[60] T. Armerding, 2016, "The OPM Breach Report: A Long Time Coming," CSO Online, October 13, https://www.csoonline.com/article/3130682/the-opm-breach-report-a-long-time-coming.html.

[61] F. Nagle, D. Wheeler, H. Lifshitz-Assaf, H. Ham, and J. Hoffman, n.d., *Report on the 2020 FOSS Contributor Survey*.

other countries cannot afford, many countries will have significant data security capabilities and none will be dominant.

Systems-Related Issues and Trends

The following sections delve into specific issues that will have a significant impact on the security of future systems and their implementation or use of encryption. In particular, some of the potential Scientific Advances discussed previously have significant Systems-related aspects that can influence the overall maturity of the products and services that are deployed to end users.

Post-Quantum Cryptography

Post-quantum cryptography refers to cryptographic algorithms believed to be resistant against adversaries possessing both large-scale quantum computers and classical computers. As discussed in Chapter 2, such algorithms are being developed and standardized today, and can be used on classical (i.e., non-quantum) computers so that systems and data are robust against future quantum computing advances. The process of transitioning to new algorithms is difficult. In mature scenarios, these issues are managed and bounded; chaotic scenarios yield insecure cryptographic implementations. The challenges that need to be addressed include:

- *Incompatibility with existing infrastructure and systems:* Post-quantum cryptography (PQC) algorithms typically involve larger secret and public keys, longer computation times, and longer message lengths. For constrained devices like the U.S. government Common Access Card (CAC) and readers, these limitations can introduce incompatibilities with existing hardware and performance challenges. Other examples include 5G cellular systems (where existing designs do not include PQC) and cryptocurrencies such as Bitcoin. All of these systems can be changed, but such updates will take time, money, and determination, in some cases to the level of complete replacement.
- *Complexity and security risks arising from the need for backward compatibility during transitions:* Because today's cryptography is so pervasive, many different protocols and implementations will need to be transitioned. During the transition phase, many ecosystems will need to run two entirely distinct security protocols or configurations in parallel—a backward-compatible option using traditional public key algorithms, and one with post-quantum algorithms. Many different components may need upgrading, including cryptographic protocols, key management hardware, certificates, and certificate issuance infrastructure. Because many devices will need to interoperate with non-upgraded systems, and some systems may never get upgraded, it may take a very long time before the legacy (non-quantum-resistant) options can be fully disabled. Transitions will likely be especially lengthy in military systems, whose lifespan is often measured in decades.
- *The potential for long-term interoperability issues arising from multiple competing systems based on national interests:* Today's Internet standards largely achieve interoperability by predominantly using a small number of NIST-standardized algorithms. In contrast, there may be multiple competing non-interoperable PQC standards—for example, one set advanced by NIST, and others advanced (or mandated) by other nations. This diversity raises a number of distinct issues ranging from technical to sociopolitical—will devices need to implement all of these protocols, will this contribute to Internet fragmentation, will critically important standards be controlled by adversaries? Protocols such as SSL/TLS that support algorithm negotiation may do better in terms of interoperability, but typically the overall protocol security is that of the weakest supported public key option. Thus, today's commercial cryptographic suites, which are already fairly chaotic, may become even more so as PQC systems are deployed and operated.

FINDING 4.12: The transition to post-quantum cryptography is likely to be prolonged over many years. It may also provide a rationale for replacing obsolete systems that have other security problems.

FINDING 4.13: The complexity of the transition to post-quantum cryptography will likely introduce a range of new security vulnerabilities.

FINDING 4.14: A new classical cryptanalysis algorithm or quantum computing development could result in rushed and disorganized efforts to replace widely used public key algorithms and cryptographic standards that use these algorithms. Such a breakthrough would require mitigation efforts that would be more complex than fixing typical software bugs, such as the coordinated deployment of major protocol updates across implementations and services.

Side Channels

Cryptographic algorithms are defined as mathematical operations. For example, NIST's standard for AES-256 encryption defines how to transform a 128-bit plaintext block and a 256-bit secret key into a 128-bit ciphertext block. From a mathematical perspective, AES-256 is believed to be extremely strong. For example, even an adversary who captures billions of plaintext/ciphertext pairs encrypted with a single random target key, and who can harness all the computers on Earth for decades, has no significant chance of recovering the key using the best-known attacks.

The situation is completely different if the adversary can gain even a tiny fraction of a bit of additional information from the encryption process. For example, cryptographic devices in practice often emit radio frequency waves, draw varying power,[62] or take variable time depending on the operations being performed. These effects are termed "side channels" because they reveal additional information beyond the mathematically defined inputs and outputs. Even very low fidelity measurements of such side channels can be enough to turn the adversary's problem from a wildly infeasible calculation into a trivial one.

Beginning in 2018 with the discovery of classes of vulnerabilities referred to as Spectre[63] and Meltdown,[64] side channels in modern processor hardware became a subject of increased interest to vulnerability researchers. As with cross-site scripting errors 10 to 15 years before,[65] vulnerability researchers reported an array of new vulnerabilities in a relatively short period. While some of these side-channel vulnerabilities can be mitigated in software or microcode updates, others cannot—and there is a real possibility that even more serious unfixable hardware vulnerabilities will be uncovered in the future.

For security engineers, side-channel attacks are challenging because they involve implementation properties that cross teams and layers of abstraction. For example, a cryptographic algorithm may be implemented in software, which runs on a CPU, which is manufactured from transistors, which are connected by wires that generate RF signals, which ultimately expose the secret key. Software developers generally have little idea of the analog characteristics of the circuits, while circuit designers generally have little understanding of the cryptographic implications of their designs. To exploit the resulting analog side channels, attackers require physical proximity to the device or need some other way to collect the required information. Relatively advanced mitigations are present in some specialized cryptographic chips such as those used in smart cards, but most processors today contain no protections.

---

[62] J. Kocher, J. Jaffe, and B. Jun, 1999, "Differential Power Analysis," *Advances in Cryptology—Crypto 99 Proceedings* (M. Wiener, ed.), Lecture Notes in Computer Science, Vol. 1666, Springer-Verlag.

[63] P. Kocher, J. Horn, A. Fogh, D. Genkin, D. Gruss, W. Haas, M. Hamburg, M. Lipp, S. Mangard, T. Prescher, M. Schwarz, and Y. Yarom, 2019, "Spectre Attacks: Exploiting Speculative Execution," 40th IEEE Symposium on Security and Privacy (S&P'19), https://spectreattack.com/spectre.pdf.

[64] M. Lipp, M. Schwarz, D. Gruss, T. Prescher, W. Haas, A. Fogh, J. Horn, S. Mangard, P. Kocher, D. Genkin, Y. Yarom, and M. Hamburg, 2018, "Meltdown: Reading Kernel Memory from User Space," 27th USENIX Security Symposium (USENIX Security 18), https://meltdownattack.com/meltdown.pdf.

[65] Symantec, 2008, "Symantec Internet Security Threat Report Trends for July–December 07," Volume XIII, April, https://web.archive.org/web/20080625065121/http://eval.symantec.com/mktginfo/enterprise/white_papers/b-whitepaper_exec_summary_internet_security_threat_report_xiii_04-2008.en-us.pdf.

In addition to analog side channels, timing side channels can also expose sensitive information.[66] These can result from cryptographic algorithm implementations that, for example, allow secret intermediate values to affect a memory address or conditional branch (enabling Flush+Reload attacks[67]). In addition, packet timing and sizes can reveal information about network flows as well as messages that cannot be decrypted. Modern microprocessors also have numerous side channels internally that can leak information between processes. In general, optimizations by chipmakers to increase CPU performance tend to introduce new side channel vulnerabilities.[68] As a result, new side channel attacks are likely to be discovered.

Although side-channel vulnerabilities and attacks draw academic and media attention, their real-world impact may be less than that of software vulnerabilities that may be easier to discover and exploit. For example, many side channel vulnerabilities are only practical to exploit if adversaries can run software on the target machine, in which case software bugs are probably easier to exploit. Researchers will likely continue to discover side channel vulnerabilities, and end users, including the Intelligence Community will be well advised to assess their potential impact and respond accordingly.

5G Cellular Systems

5G cellular systems are now being built out around the world; although their cryptography is not particularly novel in its approach, 5G systems as a whole deserve considerable attention from the Intelligence Community.[69] Most press coverage of 5G systems has emphasized its new radio technology—for example, the increased bandwidth that can be provided via millimeter wave technology. Although such radio issues may have some relevance for the Intelligence Community—for example, radio intercepts will be more difficult with many short-range radio links, distributed beam forming, and so on—they are largely independent of cryptographic considerations.

For many use cases, the 5G network is viewed as untrusted with respect to privacy and integrity of communications. An additional layer of cryptographic protection (such as SSL/TLS for web services or VPNs) provides protection unrelated to the network. However, the network still sees traffic flow metadata, including protocol information, packet sizes and timing, and source and destination addresses. In addition, network vulnerabilities can also result in denial-of-service attacks. These present the greatest systems-level concerns, as they cannot be addressed easily via application layer cryptography. Current networks also collect fine-grained location information about users; the 5G standards aim to provide more privacy.

One of 5G's more interesting aspects lies within its conversion to a distributed cloud infrastructure, in which services are implemented almost entirely in software. For example, Verizon's 5G "Mobile Edge Cloud" service is provided by Amazon Web Services hosted within nearby (low-latency) data centers.

Older cellular systems generally provided services (echo cancellation, texting, voicemail, etc.) via suites of specialized hardware and software appliances provided by equipment manufacturers (Huawei, Ericsson, etc.). The 5G vision sees such services provided by chains of virtual machines running within "edge clouds." The edge cloud vision, sometimes identified by the dense acronym NFV/SDN, for Network Functions Virtualization and Software-Defined Networks, or C-RAN for Cloud Radio Access Networks, moves to an architecture based on software running on commodity servers. In its most extreme form, some of the radio functionality is also envisioned as running in such edge clouds. Some nations envision these software suites as fully open source rather than provided by a traditional vendor.

---

[66] P.C. Kocher, 1996, "Timing Attacks on Implementations of Diffie-Hellman, RSA, DSS, and Other Systems," pp. 104–113 in *CRYPTO 1996*.

[67] Y. Yarom and K. Falkner, 2014, "Flush+Reload: A High Resolution, Low Noise, L3 Cache Side-Channel Attack," pp. 719–732 in *Proceedings of the 23rd USENIX Security Symposium*, https://www.usenix.org/system/files/conference/usenixsecurity14/sec14-paper-yarom.pdf.

[68] P. Kocher, J. Horn, A. Fogh, D. Genkin, D. Gruss, W. Haas, M. Hamburg, M. Lipp, S. Mangard, T. Prescher, M. Schwarz, and Y. Yarom, 2019, "Spectre Attacks: Exploiting Speculative Execution," 40th IEEE Symposium on Security and Privacy (S&P'19), https://spectreattack.com/spectre.pdf.

[69] P. Marsch, Ö. Bulakci, O. Queseth, and M. Boldi, eds., 2018, *5G System Design: Architectural and Functional Considerations and Long Term Research*, John Wiley & Sons, Hoboken, NJ.

This evolution is likely to introduce two very significant systems issues in practice. First, the software will likely be highly complex and thus introduce many possibilities for error. In addition, the perceived ease with which software can be updated may result in poor implementation quality, as vendors may prioritize speed in deploying and improving implementations over making them reliable or secure. If so, the rate of software upgrades may accelerate considerably. In such scenarios, when one looks across the world, one might see many different software versions (and bugs) running in the various providers rather than today's relative uniformity.

Second, telecommunications providers may find it difficult to manage such software-intensive, cloud-based infrastructure, particularly when their infrastructure spans a number of distributed edge-clouds; staffing will likely be an issue. Their difficulty will increase even more to the extent that they migrate to an agile "DevOps" (software development [Dev] and IT operations [Ops]) style of operation, which is radically different from their current cultures. It seems unlikely that cellular providers in most parts of the world will be able to support such infrastructure development and upgrading without considerable help from third parties, thus, in essence outsourcing their operation (and the correct functioning of their cryptographic systems). The combination of new architectures, complex and evolving software, operator inexperience, and perhaps outsourced operation may elevate the chaos, rather than maturity, within many provider networks.

Thus, today's cellular infrastructure, which is relatively mature, may in coming years become more chaotic for some service providers as they roll out their 5G infrastructure.

FINDING 4.15: 5G may introduce a number of new systems issues in practice, owing to both complex new suites of software and operator inexperience in distributed cloud environments.

Internet of Things

The term "Internet of Things" (IoT) spans an extremely wide variety of devices and systems that range, according to context, from the Supervisory Control and Data Acquisition (SCADA) systems that oversee entire electrical grids through instrumented aircraft fleets all the way down to doorbells and thermostats. The examples in this discussion are concerned primarily with what might be termed "personal" IoT devices, such as smartphones, personal medical devices, household appliances, and automobiles.

Industrial IoT or SCADA systems are well established worldwide and share many cybersecurity considerations with IT systems. SCADA systems do pose significant unique security challenges, especially resulting from the fact that such systems are often performance critical, long-lived, and difficult to update. The cybersecurity of SCADA systems has been the subject of many studies and standards efforts—see, for example, the National Academies' report *Enhancing the Resilience of the Nation's Electricity System*[70] or International Electrotechnical Commission (IEC) Standard 62443.[71] In contrast to SCADA and industrial IoT systems, emerging consumer IoT devices are becoming ubiquitous and much less likely to be designed with security as a major consideration.

The continuing move to connected devices means that it will soon be almost impossible to live a "normal" life without being surrounded by devices with sensors, multiple communication channels (WiFi, 5G, Bluetooth) all built to be cheap, which implies little security, and opaque supply chains. These devices pose a serious risk of surreptitious data collection. For example, the cryptographic strength of an encrypted telephone system call is irrelevant if a compromised IoT device can record the audio. Likewise, cellular data capabilities in IoT devices can allow attackers to mount attacks and exfiltrate data without being seen by local network monitoring tools. Given such omnipresent, constant surveillance, implemented by poorly secured devices, it may prove extremely difficult to keep many personal matters private in coming years—such as where one has been over the past year, who one might have met, what they look like, and the substance of conversations.

---

[70] National Academies of Sciences, Engineering, and Medicine, 2017, *Enhancing the Resilience of the Nation's Electricity System*, The National Academies Press, Washington, DC, https://doi.org/10.17226/24836.

[71] International Electrotechnical Commission, 2021, "Understanding IEC 62443," blog, February 26, https://www.iec.ch/blog/understanding-iec-62443.

Although robust open-source reference implementations might reduce the amount of security expertise needed by IoT vendors, the pressures for low cost and rapid release of such devices present major challenges. In addition, even if security vulnerabilities are found and subsequently fixed for future devices, it may not be possible to apply patches to devices in the field, or users may decline to update. Even in otherwise well-secured devices, metadata may undesirably leak information.

The fact that IoT devices can kill people is an even bigger challenge. There are many warning signs. For instance, in 2019 the FDA issued an alert on insulin pump security, warning that some devices could be hacked and remotely controlled which "could allow a person to over deliver insulin to a patient, leading to low blood sugar (hypoglycemia), or to stop insulin delivery, leading to high blood sugar and diabetic ketoacidosis (a buildup of acids in the blood)."[72] Automobiles can also be very dangerous. As early as 2015, Fiat recalled 1.5 million vehicles after a demonstration of (benign) hackers taking remote control of a Jeep Cherokee through the Internet.[73] Self-driving automobiles clearly pose further potential vulnerabilities.

Many IoT issues may have unexpected but significant consequences. As one example taken from industrial IoT, in the 2013 Target breach, the attackers used remotely accessible networked heating, ventilation, and air conditioning (HVAC) equipment as an entry point into the company's network in order to install malware on point-of-sale terminals. As a second example, security vulnerabilities in poorly secured networked smart meters could potentially be escalated to destabilize the entire electrical grid through surges or blackouts. Last, it is a well-known problem that medical devices often run unpatched versions of ancient software like Windows XP (and installing updates may require FDA approval); ransomware attackers have managed to exploit vulnerabilities in these legacy devices as well as in the remote access and network firewall infrastructure intended to protect networked hospital equipment.

FINDING 4.16: Many IoT components are poorly secured and easy to subvert, with an extremely wide range of consequences that are difficult to predict but potentially very high impact for the Intelligence Community and broader society. Because IoT will likely bring significant improvements to many aspects of life, however, more money and energy may be devoted to securing such devices going forward.

---

[72] Food and Drug Administration, 2019, "FDA Warns Patients and Health Care Providers About Potential Cybersecurity Concerns with Certain Medtronic Insulin Pumps," FDA News Release, June 27, https://www.fda.gov/news-events/press-announcements/fda-warns-patients-and-health-care-providers-about-potential-cybersecurity-concerns-certain.

[73] A.M. Kessler, 2015, "Fiat Chrysler Issues Recall Over Hacking," *The New York Times*, July 24, https://www.nytimes.com/2015/07/25/business/fiat-chrysler-recalls-1-4-million-vehicles-to-fix-hacking-issue.html.

5

Scenarios

As discussed in Chapter 3, the objective of the committee was to select three scenarios that would be stressful for planning and pose a wide range of situations, as well as meet criteria derived from the focal question. Although each of the drivers is important in developing the scenarios, the committee determined that exploring the edge cases presented by the Systems (Mature/Chaotic) and Scientific Advances (Disruptive/Predictable) drivers in combination with a "Fragmented" Society and Governance driver would provide the most value to the Intelligence Community. Choosing these scenarios allowed members to explore two technology breakthroughs that would impact existing public key cryptographic algorithms: a breakthrough in quantum computing and the discovery of a new cryptanalytic method usable on classical computers. These scenarios also provided opportunities to highlight the important role that systems and cybersecurity will play in the future. Last, the committee selected three scenarios with a "Fragmented" Society and Governance driver, because such scenarios presented the greatest challenge to the Intelligence Community and, the committee agreed, seemed the more likely direction in which global events were trending. "Fragmented" scenarios also are more complex as they do not assume homogeneity with respect to legal, cultural, privacy, and consumer issues, and selecting "Fragmented" scenarios pushed members to consider unique regional perspectives and uses.

Table 5.1 shows the endpoints associated with the eight possible scenarios, and Appendix C provides a brief summary description of each. The following sections describe each of the three scenarios—Scenarios 2, 5, and 6—that the committee selected for in-depth exploration. In each case, the scenario description is presented as a historical "look back" from a future perspective. Each contains a table that summarizes the scenario and a description of the scenario in detail.

SCENARIO 2: A BRAVE AND EXPENSIVE NEW WORLD

In 2038, China announced the availability of a large-scale, commercial quantum computer service. The Chinese intelligence services have supported this development and "tested" it for several years on encrypted information of domestic and foreign targets. Several other advanced nations, including the United States, are understood to have similar capabilities but have not advertised the fact. (See Box 5.1.)

Although China was first to announce the breakthrough, the pace of scientific and technological advances throughout the previous decades made the possibility seem increasingly likely, and a breakthrough appeared imminent. A series of advances had resulted in commercially available noisy intermediate-scale quantum (NISQ) computers. These machines were optimized for chemistry simulations, and mostly marketed to pharmaceutical and biotech companies for drug discovery and agricultural applications. After an algorithmic advance that allowed a practical speedup for deep learning algorithms on NISQ computers, the major cloud computing platforms began offering time on quantum computing coprocessors. The early commercially available computers had error rates that were too high to run Shor's algorithm to break public-key encryption, but this was no matter: the market was much more interested in large-scale commercial applications. However, once several tech companies (IBM, Microsoft, Intel, and a handful of start-ups) began competing to sell quantum computing services on the commercial market, there was a rapid improvement in error rates and scaling behavior. It seemed clear that arbitrary scaling could be reached within a few years.

Because quantum computing that would threaten public-key encryption seemed like such a real possibility, however, companies and countries invested time and resources to understand how to leverage the opportunities and prepare to mitigate the risks. All of the talk about quantum computing drove a sense of urgency in the cryptographic and information technology communities, forcing many to realize that the threat to public-key cryptography was no longer theoretical and it was time to complete the transition to post-quantum cryptography (PQC). Scenario 2, "A Brave and Expensive New World," is described in Table 5.2.

Developers and architects of systems regrouped from a turbulent time. During the early 2020s, cyber-attacks on physical infrastructure and information continued to increase. It did not matter whether these were nation states, criminal enterprises, or individuals—they were disruptive, costly, and dangerous. Over the decade, countries and companies invested heavily in improving cybersecurity, focusing on high quality standards and implementations. Pressures by concerned commercial customers and individuals added to those imposed by major powers' governments (such as U.S. Executive Order 14028) and drove technology vendors and large users to follow through on achieving improved security. Fewer bugs in systems made vulnerabilities progressively rarer and nations found it increasingly difficult to undermine the confidentiality, integrity, or availability of adversaries' information systems. Mature systems and strong cybersecurity made the transition to PQC on par with past (and technically simpler) transitions. There were challenges, but, for the most part, the move happened with little global disruption.

The move to PQC was, for the most part, an orderly transition, but the changing geopolitical landscape over the decade led to a fracturing of standards and the rise of national algorithms (i.e., country-specific cryptographic technologies) with far-reaching consequences. The same sense of urgency that drove the transition to PQC provided an impetus for countries to start developing and implementing their own standards. For example, the National Institute of Standards and Technology (NIST) in the United States and its counterparts in China and Europe each developed standards that were accepted by the International Organization for Standardization. In the early 2030s,

TABLE 5.1 Endpoints That Define the Eight Possible Scenarios

Scenario	Scientific Advance	Society and Governance	Systems
1	Predictable	Fragmented	Mature
2	Disruptive	Fragmented	Mature
3	Predictable	Global	Mature
4	Disruptive	Global	Mature
5	Predictable	Fragmented	Chaotic
6	Disruptive	Fragmented	Chaotic
7	Predictable	Global	Chaotic
8	Disruptive	Global	Chaotic

TABLE 5.2 Scenario 2: A Brave and Expensive New World

Scenario Title	Driver Endpoints	Highlight
A Brave and Expensive New World	Disruptive Fragmented Mature	A breakthrough in quantum computing is balanced with more secure systems and software and an orderly transition to post-quantum encryption.
Scenario Description		
This scenario posits that a breakthrough in quantum computing is offset by an orderly transition to post-quantum encryption and other emerging cryptographic techniques, because of earlier investments in systems and cybersecurity. Overall, the balance now favors defense. However, the global political picture remains fragmented. The bottom line for the Intelligence Community is that offensive cryptography efforts have become more difficult, and the alliance structure that is a major plus for U.S. intelligence is less reliable and more fluid. In this scenario, a major issue for U.S. intelligence will be its ability to discern, far enough in advance, the development of a reliable, large-scale quantum computer. This is of crucial importance to this scenario, which posits the development of more secure systems because governments and the private sector take seriously the need for improved cybersecurity and invest in the development and deployment of much more robust systems. Even if such a quantum computing breakthrough did not occur, there are obvious benefits in enhanced systems and cybersecurity. If these steps are taken early enough and on a wide basis, then the transition to a post-quantum world would be orderly. It will be important for U.S. intelligence to be able to discern in advance the development of a reliable large-scale quantum computer both to ensure that progress toward post-quantum encryption is sufficient for defensive purposes and to assess potential offensive opportunities. International relations in this scenario fragment into a small number of blocs composed of a few major powers (the United States, the European Union, China) and a large number of dependent powers that, for the most part, rely on a major power for economic, technological, and defense support. Technology and encryption are similarly fragmented with blocs sharing common approaches to encryption and inter-bloc communications relying on weak "least common denominator" encryption. Within blocs, government surveillance is the norm although some governments and some private companies use new capabilities for processing encrypted data to protect sensitive processing.		

Europe, led by Germany's BSI,[1] decided that, although NIST ran a good process, they had selected the wrong winners; therefore, the European Union (EU) determined that it would select a different set of winners. These "Big Three" powers saw opportunities for influence and, rather than align on a single standard, each major power forced vendors to implement by default and individual and corporate users to adopt their own preferred national suite of encryption algorithms.

This fragmentation continued over the intervening years, and the "Big Three"—China, Europe, and the United States—have emerged as major powers exerting regional and digital influence over their own blocs of loosely allied nations. Some countries have plotted their own course, for the most part, and move between the major spheres of influence, depending on where their interests lie—the fragmented nature of global politics means that alliances are constantly shifting. In the world of 2040, the United States finds that, rather than a stable of trusted allies, it has developed a series of relationships based on the situation and the interests at hand.

Shifting alliances and diminishing trust compared to the 2020s and before mean that, at this point, there is very little globalism; instead, the "Big Three" blocs concentrate trade and technology within their own systems. Each major power has developed and deployed its own information technology systems, quite distinct from the others, with its own hardware designs and implementations, software, and development processes. Each also has its own mobile wireless system, its own fiber optic networks and technology, its own extensive low-Earth orbit communication satellite systems, its own Global Positioning System equivalent, and its own integrated infrastructure for autonomous transportation (air and ground). After a number of physical and cyber infrastructure attacks in the early 2020s, the "Big Three" invested significant resources in building solid defenses into their information systems; in fact, the systems of 2030 are much more secure than the systems of a decade ago—they simply do what they are designed to do.

Global technology firms do not exist as they did a decade ago. In the United States, firms such as Meta, Amazon, and Google could not overcome the backlash from questions of privacy and disinformation, and Congress had stepped in to regulate them heavily. For different reasons, government and regulatory authorities in Europe and China took similar steps, encouraging the rise of national champions that would control the flow of information

---

[1] Bundesamt für Sicherheit in der Informationstechnik: German Federal Office for Information Security.

> **BOX 5.1**
> **Description of This Notional World**
>
> **What If ... a Nation State Had Scalable Quantum Computing Capabilities Years Before Admitting Its Existence?**
>
> In general, the committee agrees that scalable quantum computing will be difficult to develop in secret. The investment of time and resources would be considerable and a challenge for any organization or country to keep hidden, especially in a world of international collaborations on quantum computing research where the absence of a country's leading researchers from publications and conferences would be very noticeable. But history has shown that, given enough resources and a focused mission, large projects can be kept secret long enough to have a significant impact on global affairs and conflict. The Manhattan Project and the development of the first stealth aircraft are examples of the ability to do so, even in an open society.
>
> The committee believes it is important for the Intelligence Community to engage in imaginative thinking about "What If??" What if the United States were to develop a scalable quantum computer in secret? How might the United States use such a capability legally and ethically? How would we keep our capability secret and decrypt our adversaries' communications? It is easy to make comparisons to Arlington Hall, Bletchley Park, and the codebreaking efforts undertaken there, but that was during a period of open conflict with a focused target. How might the U.S. Intelligence Community leverage the ability to decrypt almost any encrypted data? What would the long-term implications be when the capability, inevitably, came to light? How might that affect the trust that citizens or allies have in the U.S. government?
>
> At the same time, the Intelligence Community should explore the question from the defense perspective. What if the tables were turned and a competitor were to build scalable quantum computing in secret? What are the indicators and warnings that might signal a breakthrough is imminent? What researchers', laboratories', or industries' activities might provide indications of such a development? How should the Intelligence Community treat communications if they think an adversary has made such a breakthrough?
>
> Although these questions have no easy answers, exploring them will help the Intelligence Community prepare for the possibility of a breakthrough and consider appropriate steps.

within their spheres of influence. Russia and India have attempted to build their own competing systems, but their influence and reach do not compare to those of the "Big Three" major powers.

Dependent nations rely on the national algorithms from a major power for their international banking systems, their information technology, and so forth. In most cases, they adopt the cryptographic systems of their major power partner but are generally suspicious that there may be backdoors or weaknesses in such systems and are certain that they are not getting "the really good stuff." They are also heavily reliant on the "Big Three" for vaccines and other bio-engineering products; new diseases are sweeping through all the time, endangering both people and agriculture. There is much speculation that these diseases are man-made, but no proof. In any case, most nations rely on a major power for protection and to help arbitrate the international issues driven by climate change such as shortages of water and mass migrations.

Trade among blocs exists, although the underlying communications infrastructure is something of a mess. Because each bloc requires support for its own cryptographic algorithms, vendors simply implement support for a long list of algorithm variants that are all considered to be roughly reasonable choices. International corporations maintain redundant parallel internal information technology (IT) infrastructure in order to satisfy the complex regulatory environments of different regions and major powers. EU privacy regulation has begun requiring that customer records be stored and operated upon in entirely encrypted form when possible. This has resulted in commercially available multi-party computation (MPC) and encrypted database search platforms with sufficient performance and usability to be viable alternatives to their unencrypted counterparts for some applications. The technical details and privacy guarantees of these schemes are widely misunderstood by governments and end users.

Although the complexity of true fully homomorphic encryption (FHE) has not improved enough to allow its use for nontrivial systems, the acronym FHE has ended up being used by nontechnical users as a generic term applying to just about any privacy-preserving system. Several traditionally democratic countries have mandated law enforcement access to encrypted communications, with some also mandating cryptographic schemes for accountability and tracing for all electronic speech. The particular implementations have been left up to service providers. Citizens assume that intelligence agencies regularly gain access to the law enforcement access portals to surveil communications at will. The major encrypted communication protocols all include complex cipher negotiation steps; a few downgrade attacks are published in the academic literature, but mature systems development has limited the real-world impact of such attacks.

If there is one thing that people around the world share in this scenario, it is their discontent. They are deeply suspicious and prone to expect the worst from their own governments and their fellow citizens, but the manifestations are quite different in the blocs. The United States has long attempted to rally support from its own citizens by emphasizing transparency; other nations have implemented extremely intrusive "thought control" regimes by intense monitoring of all citizens, backed up by extensive use of artificial intelligence to collect and analyze data. Even in the European Union, despite a historical commitment to human rights, a rise in terrorism and threats posed by massive illegal immigration from Asia and Africa led to increased efforts to monitor and control the population. Whatever the approach, the end results are similar: a large majority of citizens feel that their government is controlled by corrupt elites who are trying to suppress them and view their fellow citizens with deep distrust.

One additional dynamic is at play. Many of the Major Powers have aging populations where a rapidly shrinking workforce must support an ever-larger contingent of retirees. Of course, artificial intelligence and robotics help in this regard, but this trend brings two big societal strains that threaten the national and international order.[2] Within a Major Power, younger generations are becoming ever more frustrated with their incomes being transferred to ever-larger numbers of relatively affluent retirees. And within the larger bloc, the dependent nations are on the whole much younger—and more dynamic—than the corresponding Major Power. Both strains tend to undermine the Major Power government, leading to a pervasive sense of unfairness and instability for all Major Powers.

Warning Signals for Scenario 2

Table 5.3 summarizes some key developments associated with Scenario 2, along with warning signals that might indicate to the Intelligence Community that this scenario is emerging.

Risks and Opportunities in Scenario 2

"A Brave and Expensive New World" appears to be great powers jockeying for territory and influence across a series of constantly changing alliances. But in this scenario, technology has advanced to the point where territory has been replaced by information as a symbol of power. Indeed, companies that are influenced by great powers or companies that wield great power and influence in their own right because of their technological advantages also have to be considered as significant players in this scenario. The combination of Disruptive, Fragmented, and Mature will present its own unique challenges for the Intelligence Community. For example, in a Fragmented scenario, nations are less likely to trust each other or share secrets and resources. Looking more locally, the Intelligence Community might face the challenge of recruiting, hiring, and retaining high-quality staff. With mature systems being pervasive (as described by the endpoint on the Systems driver), accessing information will have to rely less on technology and exploiting bugs and more on overt surveillance and human intelligence and exploiting people—presenting an insider threat challenge for intelligence services across the globe.

Pervasive mature systems would have deep, far-reaching effects across society. In this scenario, cybersecurity is very high quality and breaks are rare. Consequently, nations find it very difficult to gather intelligence from, or disrupt, information systems. This would be a game changer for the Intelligence Community.

---

[2] See, for example, B. Nichiporuk, 2000, *Demographics and the Changing National Security Environment*, RAND Corporation, Santa Monica, CA.

TABLE 5.3 Key Developments and Warning Signals for Scenario 2: A Brave and Expensive New World

Development	Warning Signal
"A series of advances had resulted in commercially available noisy intermediate scale quantum (NISQ) computer."	The Intelligence Community should be paying attention to the commercial world and who is developing a NISQ and how quickly.
"The major cloud computing platforms began offering time on quantum computing coprocessors."	Is quantum computing being offered as a service? Even if it is not scalable, the Intelligence Community should be monitoring who is selling and who is buying these services.
"During the early 2020s, cyberattacks on physical infrastructure and information continued to increase. It did not matter whether these were nation states, criminal enterprises, or individuals—they were disruptive, costly, and dangerous. Over the decade, countries and companies invested heavily in improving cybersecurity, focusing on high quality standards and implementations."	Are there more attacks over the next few years; are more high-quality standards and (especially) implementations becoming available? If so, this world might be emerging; if not, we might be on a Chaotic Systems axis.
"But the changing geopolitical landscape over the decade led to a fracturing of standards and the rise of national algorithms (that is, country-specific cryptographic technologies) with far-reaching consequences."	Are national standards and encryption algorithms emerging around the world or are countries continuing to engage in and adopt international standards?
"In the United States, firms such as Facebook, Apple, and Google could not overcome the backlash from questions of privacy and disinformation, and Congress had stepped in to regulate them heavily. For different reasons, government and regulatory authorities in Europe and China took similar steps, encouraging the rise of national champions that would control the flow of information within their spheres of influence."	Is Congress regulating the major technology companies more heavily? Are Europe and China taking similar steps?

Risks for the Intelligence Community in Scenario 2 include the following:

- *Cryptographic breakthroughs pose significant risk:* Although there was not a quantum breakthrough until the late 2030s, the emergence of quantum computers means that information that was encrypted using public-key encryption algorithms vulnerable to quantum computer attacks and collected and stored by an adversary is vulnerable to disclosure. Once such information is collected, there is nothing the originator can do to protect it, but there is a need for an inventory, assessment of potential damage, and development of mitigation plans. In addition, any data previously collected by the Intelligence Community that was secured by vulnerable encryption could be accessible.
- *Insider threat:* As distrust in the U.S. government increases, the risk from insider threat grows. In addition, given the strength of cybersecurity across the globe, the premium placed on human assets in organizations will only increase the demand for insider access.
- *Recruiting, hiring, and retaining the "best and the brightest":* Employee quality is a key risk. Will the United States and the Intelligence Community have the talent to be successful, particularly if the United States ceases to attract students and faculty from around the world?
- *Weakening and shifting alliances:* In a fragmented scenario, the United States should not assume that traditional long-term allies remain. As trust and collaboration recede, collection efforts grow more expensive and will surely need deep triage. With ever-shifting bilateral alliances, it may be hard (slow) to pivot to new targets or allies; and it is unwise to share important secrets with any nation that seems likely to switch alliances.
- *Mature implementations of advanced cryptography:* Multi-party computation may permit limited or controlled data sharing with allies, but if used by adversaries it could be an impediment to collection efforts that would historically have targeted plaintext. Similarly, mature implementations of anonymous cryptocurrencies would make following money flows much more difficult.
- *Too many targets:* If each country, or each region, is using its own unique system or form of encryption it is possible that the Intelligence Community would not have the resources to monitor and penetrate all systems

of interest. Fragmentation of the scenario's information infrastructures increases the cost of intelligence collection from information systems, as every system requires its own tailored implementation.

Opportunities for the Intelligence Community in Scenario 2 include the following:

- *The gloves come off:* Perhaps there will be fewer problems of technical blowback as globalism recedes; regionalization of systems and technology means that a different region is considered "other," as opposed to today where uniform standards and systems mean shared risk. Tools used to penetrate other nations may not also work against the United States. The Intelligence Community may also be able to exploit the seams in fragmented international systems, as the possible need to convert from one encryption protocol to another or to use weak "interoperable" systems may create exploitable weakness. In addition, the Intelligence Community may be able to use new analytic techniques against laggards in the transition to post-quantum.
- *Exploiting human assets/insider threat:* As the Intelligence Community recognizes that intelligence gains are less likely to come through attempted surveillance of adversaries' well-defended networks, the community will devote more resources to recruiting and utilizing human assets in a far broader way. This task will be facilitated mostly by the greater willingness of disaffected individuals in this scenario to work against governments they distrust or even disdain, and partly by the fact that in a fractured geo-political situation, there would be multiple opportunities (potentially within an adversary country and across a wider number of adversaries) to find potential human intelligence assets who could add value. Such assets could help penetrate otherwise difficult networks and communications, and, through the ability of human sources to report on other types of information, would help offset the intelligence loss stemming from inaccessible networks. Properly placed assets could also facilitate technical attacks on their countries' cryptographic systems by influencing national encryption standards while they were under development.
- *New opportunities:* If mature implementations become available for computation on encrypted data, this may open novel opportunities. Computation may become possible in untrusted or partially trusted environments such as other regions or allied government facilities using cryptographic techniques for "privacy-preserving transparency" (via MPC and zero knowledge).

Actions Relevant to Scenario 2

- *Move to mature systems:* This will require enormous effort across the United States, as the current status is at the "chaotic" end of the spectrum for nearly all systems. It is not clear that anyone really understands the implications of good cybersecurity because there is no real-world experience to draw on. It is also unclear how to move toward such systems in practice, although the 2021 cybersecurity executive order appears to be an attempt to move in that direction. This issue is much larger than the Intelligence Community and would require effort from major companies, researchers and start-ups, educational institutions, and the U.S. government. It is likely that very heavy investments would be required for research but especially for real-world implementations. As mature systems become available, the Intelligence Community must introduce new, more resilient IT very quickly (not over decades) and as needed abandon less-important IT infrastructure that cannot be quickly secured.
- *Focus on ensuring that the United States can provide an adequate supply of high-quality, trustworthy staff:* Although the large number of foreign-born students in science, technology, engineering, and mathematics (STEM) is a major advantage for our country, it should not be taken as a given and may not continue. The U.S. government should work toward considerable improvement in K–12 mathematics and science education, in particular. Advanced cryptographic research is poorly and unevenly funded in the United States; steady, long-term funding should be available for graduate students in cryptography and for efforts that may lead to mature systems (e.g., practical assured programming languages).
- *Invest in technologies that can enhance the U.S. posture for both offense and defense:* Potential areas of investment for defense include safer programming languages, improved toolchains, and formal verification. Other areas to consider for improving defense include realistic applications for the deployment of computing

on encrypted data and applications of trusted execution environments. On the offensive side, finding applications for noisy intermediate-scale quantum computers that could lead to a virtuous cycle in quantum computing technology could pay dividends.

- *Limit/mitigate fragmentation, both in technologies and between the United States and allies:* The Intelligence Community and the United States will need to consult with current allies and try to leverage purchasing power to push common standards and increase implementation maturity. The Intelligence Community will need to emphasize its efforts to preserve existing relationships with allies, but potentially plan to pivot toward new, perhaps short-term relationships. Such a pivot will require political resources, knowledge of other cultures, and tending such relationships. The Intelligence Community can also cooperate with other expert agencies, such as the Department of Homeland Security and NIST, to ensure appeal to the broadest audience worldwide. Last, because technology fragmentation is a key feature of this scenario, the Intelligence Community needs to learn new technologies and standards; there will be many more of them than at present, and with a higher rate of disruptive scientific advances.
- *Examine resource and funding priorities:* Collection will be much more difficult and will rely far more on human intelligence than on breaking into computer systems and intercepting communication. There is a multiplicity of hardware platforms, software, and communications systems without a central point; this combination will be very challenging, and ultimately have budgetary implications. In addition, a Fragmented scenario two decades from now means that the United States might not be able to rely on traditional alliances and will need to consider the implications for intelligence collection and resources.

SCENARIO 5: THE KNOWN WORLD, ONLY MORE SO

In 2030, the breakthrough for scalable quantum computing is far off; perhaps even more surprising is the fact that there have been few encryption-related mathematical breakthroughs globally. Both objectives have had their fair share of investment and attention, but neither quantum computing nor classical mathematics have had a breakthrough moment. Without the sense of urgency that a cryptanalytic breakthrough would drive, global society has been slow to make the transition to PQC algorithms for the past decade. Forward-leaning companies and countries have invested time and money into the effort, but the transition has been uneven across the globe, resulting in incompatibility, introducing bugs and, in some cases, creating new vulnerabilities or breaking existing communications infrastructure. Scenario 5, "The Known World, Only More So," is described in Table 5.4.

The number of Internet of Things (IoT) devices has grown exponentially over the past two decades from almost 8.75 billion in 2020 to more than 250 billion in 2030 and today, in 2040, it is estimated that for each person on the planet there are approximately 1,000 connected devices. Further complicating the issue, the transition from 5G to 6G was messy and uneven across the globe. This combination of circumstances contributed to making life easier for "offense" over the course of the decade, as there were plenty of bugs in systems to exploit, driven by a reliance on poor quality code, inconsistent support for patching of security vulnerabilities, and a lack of common standards. Threat actors recognized that a messy transition to post-quantum algorithms would present an opportunity: there would be a target-rich environment and the payoff would be commensurate with the investment required.

The lack of common standards did not come as a surprise to those watching events unfold over the past decade. With no imminent threat of scalable quantum computers or classical breakthroughs that experts had said could arrive by 2030, there was no sense of urgency to align on a single, advanced global standard. The standardization process was beset by repeated delays and technical difficulties. After NIST released its final round choices for quantum resistant algorithms, the National Security Agency (NSA) raised and then escalated through the National Security Council which approved it, a request that NIST modify the parameters for the algebraically structured lattice algorithms. However, they refused to provide a detailed public explanation, leading to an uproar and rumors of either some kind of unreleased algorithmic attack against ring-LWE or an attempted backdoor.

TABLE 5.4 Scenario 5: The Known World, Only More So

Scenario Title	Driver Endpoints	Highlight
The Known World, Only More So	● Predictable ● Fragmented ● Chaotic	With no major breakthroughs and a continued lack of focus on systems and security, breaches remain common; meanwhile, the slow pace of technology change has allowed emerging competitors the chance to "catch up."
Scenario Description		
This scenario posits that there are no major breakthroughs regarding a quantum computer, as well as a continued lack of focus on systems and security. Therefore, system breaches remain common. Also, the slow pace of technology change has allowed emerging competitors the chance to "catch up." The overall balance continues to favor offense. The key issue under this scenario for U.S. intelligence is the broadening of the threat. We are already in a world where many states and a growing number of non-state actors pose threats to U.S. and allied interests. These threats will likely increase in number and severity in this scenario, potentially compounded by weakening of traditional alliances and partnerships. At the same time, a world of this sort remains a "target rich" environment for U.S. intelligence collection efforts. The constantly growing Internet of Things (IoT) also adds to this threat and to the opportunity. There is also the question of what tools, if any, the Intelligence Community should develop to respond to/retaliate against such attacks. Policies and decisions to do so belong to the policy community but the Intelligence Community should be prepared to offer a range of responsive options. Success in this scenario requires a very serious assessment of U.S.—meaning government at all levels and the private sector—vulnerabilities, in order to take steps to ameliorate or eliminate them. Given that this is the world we largely know at present, it may be difficult to motivate people and organizations to make changes that appear to be expensive and time consuming for perhaps modest gains in security. Private sector firms may also be wary, if not suspicious, about government efforts to foster changes. The measures in the Biden administration executive order are largely voluntary. Legislation may be necessary (like seat belt and speed limit laws) to foster real change. The Intelligence Community could be asked to give advice as to what measures are needed, although this sort of participation by U.S. intelligence is likely to foster greater suspicion about backdoors or other means of government intrusion.		

In response to these rumors, the European Union launched its own "independent" algorithm standardization process, which itself was beset with controversy after a set of leaked emails suggested that the Brainpool[3] elliptic curve parameters had been manipulated by Germany's BSI to permit a previously unknown trapdoor of unknown cryptanalytic impact. A segment of the cryptographic community retaliated by boycotting both contests and organizing a non-governmentally affiliated algorithm contest. This resulted in a mess of competing cryptographic standards and recommendations: NIST, an EU standard that eventually adopted the original lattice parameters that had been modified by NSA, and the grassroots cryptographic effort settled on a variant of Supersingular Isogeny Key Encapsulation (SIKE) and hash-based signatures that had widespread support among the tech industry.

The Internet Engineering Task Force was mired in discussions for years about how to incorporate algorithm negotiation into Transport Layer Security (TLS) and other higher-level protocols; the world was left with an officially supported set of TLS cipher suites that pointedly excluded the modified NIST algorithms, as well as Request for Comments–documented unofficial TLS extensions for the NIST algorithms and a number of other regionally standardized algorithms. This messy debate slowed the adoption of post-quantum algorithms and protocol upgrades by more than a decade.

In the meantime, the controversy over both classical and PQC algorithms as well as the general fragmentation of technology led to an increasingly complex regulatory and cultural situation with respect to cross-border network communications. Only vestiges of the open web, email, and other network protocols that dominated the history of the public Internet remain. Instead, communications and information have been mediated largely through individual apps. Users have become accustomed to installing custom "region" drivers and virtual private network (VPN) software that install the national algorithms, local government root certificates, and other cryptographic parameters required to access network resources in a given country.[4]

---

[3] M. Lochter and J. Merkle, 2010, "Elliptic Curve Cryptography (ECC) Brainpool Standard Curves and Curve Generation," Independent Submission to the Internet Engineering Task Force, Request for Comments: 5639, March, https://www.ietf.org/rfc/rfc5639.txt.

[4] Krebs on Security, 2021, "Adventures in Contacting the Russian FSB," June 7, https://krebsonsecurity.com/2021/06/adventures-in-contacting-the-russian-fsb.

International standards in 2030 were not the only area challenged by fragmentation. The longstanding intelligence relationships, such as the "Five Eyes," on which the United States traditionally has relied, were not as productive as they were a decade ago. Digital threats to business, supply chain, and infrastructure continued to increase throughout the decade and, by 2030, the United States and its allies had found themselves on different sides in a number of regional conflicts. Although traditional force-on-force conflict had diminished over the decade, many countries followed the model pioneered by the Iranians, North Koreans, and Chinese in the early part of the decade and invested in offensive digital and cyber technology, training, and tactics. The lack of scientific breakthroughs and engineering progress in cybersecurity meant that many countries once considered to have "inferior" militaries found they could have an impact and be disruptive on a global stage. Where the United States had traditionally been able to focus much of its attention on a few major players, it became clear by 2030 that, owing to digital vulnerabilities and the ease of creating significant mischief, almost any country, or even cross-border criminal gangs and skilled individuals, had to be regarded as a serious threat to national security. The rise in various forms of cybercrime, including ransomware and fraud, continued to challenge the ability for national and international law enforcement to keep pace. A challenging economic landscape in many parts of the world meant that cyber cartels were able to take advantage of a technically sophisticated population and put them to work. In such an environment, with multiple and shifting adversaries, it became challenging for the United States to have a coherent policy to deter and respond to such mischief and threats.

With the continued lack of attention paid to cybersecurity over the decade, most efforts to break into systems were at least moderately successful and rarely detected before countries or criminals could achieve their aims. Companies and governments are under a constant barrage of attacks across the globe. Major IT companies drag their heels on compliance with a growing tangle of Federal regulations; as just one example, they still have not implemented the exceptional access mechanisms that permit authorized parties to unlock phones and other devices for law enforcement purposes. Most citizens agree with the companies' positions as they fear an overly intrusive Federal government; university students not only actively demonstrate against such mechanisms and initiatives but many refuse to serve in the military or join the Intelligence Community. The torrent of data from the IoT has essentially obliterated privacy. (Of course, intelligence agencies around the world purchase as much personal data as they can afford. To some extent they do not need extensive collection capabilities as industry is doing it for them.) There have been proposals that consumers could "opt out" of such intrusive collection by paying a monthly service fee, but at present this seems unlikely.

Warning Signals for Scenario 5

Table 5.5 summarizes some key developments associated with Scenario 5, along with warning signals that might indicate to the Intelligence Community that this scenario is emerging.

Risks and Opportunities in Scenario 5

Although "The Known World, Only More So" might appear similar to "The World Today in 2030," the combination of Predictable, Fragmented, and Chaotic presents a challenging and complex operating environment for the Intelligence Community. In this scenario, there is no quantum computing or classical cryptographic break driving a sense of urgency or advantage/disadvantage, and that gives a number of global competitors (countries and non-state actors) a chance to catch up. In addition, the combination of Fragmented and Chaotic will lead to a constantly contested information environment.

Risks for the Intelligence Community in Scenario 5 include the following:

- *Challenge for defense:* Buggy implementation of PQC combined with an uneven rollout will create weak software systems—perhaps weaker than if the post-quantum transition had not been attempted.
- *Deteriorating or diminished partnerships:* Fewer national alliances, and those that survive will be more transactional or issue focused.

TABLE 5.5 Key Developments and Warning Signals for Scenario 5: The Known World, Only More So

Development	Warning Signal
"Without the sense of urgency that a cryptanalytic breakthrough would drive, global society has been slow to make the transition to post-quantum cryptographic algorithms for the past decade. Forward-leaning companies and countries have invested time and money into the effort, but the transition has been uneven across the globe, resulting in incompatibility, introducing bugs and, in some cases, creating new vulnerabilities or breaking existing communications infrastructure."	With no cryptanalytic breakthrough (quantum or traditional) is the Intelligence Community seeing a messy and slow transition to post-quantum encryption?
"Further complicating the issue, the transition from 5G to 6G was messy and uneven across the globe. This combination of circumstances contributed to making life easier for 'offense' over the course of the decade, as there were plenty of bugs in systems to exploit, driven by a reliance on poor quality code, inconsistent support for patching of security vulnerabilities, and a lack of common standards."	Is the transition from 5G to 6G messy and are threat actors taking advantage of the "messiness"? Is offense gaining or maintaining an advantage over defense?
"In response to these rumors, the EU launched its own 'independent' algorithm standardization process, which itself was beset with controversy after a set of leaked e-mails suggested that the Brainpool elliptic curve parameters had been manipulated by Germany's BSI to permit a previously unknown trapdoor of unknown cryptanalytic impact."	Is the European Union (or any other similar major government) appearing to prioritize its own standardization process and deemphasize international standards?
"Only vestiges of the open web, e-mail, and other network protocols that dominated the history of the public Internet remain. Instead, communications and information have been mediated largely through individual apps. Users have become accustomed to installing custom 'region' drivers and virtual private network (VPN) software that installed the national algorithms, local government root certificates, and other cryptographic parameters required to access network resources in a given country."	Do the open web and Internet protocols look the same worldwide, or are communications mediated through apps? To what extent are countries locking down access to the Internet and requiring users to take additional country-unique steps to access it?
"The long-standing intelligence relationships, such as the 'Five Eyes,' on which the United States traditionally has relied, were not as productive as they were a decade ago. Digital threats to business, supply chain, and infrastructure continued to increase throughout the decade. and, by 2030, the United States and its allies had found themselves on different sides in a number of regional conflicts."	How are allied relationships progressing? Are there issues that are straining traditional alliances?

- *Rise of national or regional technology requirements:* The rise of regional information networks will require international software vendors to support multiple versions of software, including encryption technologies, to meet national or regional requirements. This will present an economic and business challenge, but also force the Intelligence Community to design specific collection strategies and resources for each region and drive the cost of technical and human resource investments.
- *Narrowing lead for U.S. Intelligence Community:* Because there have been no technology breakthroughs and, hence, no sense of urgency, other competitors have had time to catch up, narrowing the advantage of U.S. intelligence. This means that smaller and smaller state and sub-state entities will have the capability to launch attacks that, at the very least, will be disruptive and require Intelligence Community attention.
- *Disinformation drives poor investments:* This scenario has no breakthroughs, but how can the Intelligence Community be certain? Rising disinformation over the 2020s poses a risk that, although intelligence assessments might not show the existence of quantum computers, policymakers may make decisions or force investments as if a foreign adversary were on the cusp of developing one.
- *Shifting scope and authorities:* As cyberattacks and data breaches are constant and on the rise in this scenario, the Intelligence Community will be under pressure to focus on protecting U.S. and allied infrastructure more broadly.

Opportunities for the Intelligence Community in this scenario include the following:

- *Offensive advantages:* A Fragmented and Chaotic scenario presents challenges for defense, but it should also present a wealth of offensive opportunities for the Intelligence Community. There will be many divergent implementations, each with its own bugs to be exploited. The lack of interoperability owing to regionalization might also lead to more plaintext or weakly protected information being accessible. In addition to more opportunities, the attacks can be more aggressive because regionalization will limit second- and third-order impacts resulting from disclosure and repurposing of U.S. offensive tools and techniques against U.S. and allied entities.
- *Take advantage of time:* In a Predictable scenario, there is time to invest in transitioning critical systems to post-quantum encryption, spending "what it takes" to overcome chaotic system development and operation before the threat becomes a crisis.

Actions Relevant to Scenario 5

- *Resource a smooth transition to post-quantum algorithms:* These algorithms should be standardized and then adopted within the Intelligence Community as soon as possible. In some cases, this will require new hardware in addition to new software—for example, PQC certificates may exceed embedded hardware memory sizes. It would be a mistake to gradually "trickle out" post-quantum algorithms; upgrades should be identified, prioritized in the budget, and done quickly.
- *Maintain trust:* International buy-in to the NIST standardization process is important for a smooth post-quantum algorithm transition timeline. As discussed in Chapter 2, this trust has been challenged in the past by reactions to the actions of the U.S. government. The Intelligence Community should tread carefully in the future to avoid further loss of trust or derailing of goodwill, both of the U.S. public and the international community.
- *Take steps to unify the global tech industry around a common core of algorithm and protocol standards:* NIST is best positioned to lead this effort from within the U.S. government, but the Intelligence Community can play a very important role both in bringing cryptographic expertise to bear, and in strongly encouraging American industry to prioritize such actions. To the extent that government requirements are practical and align with the needs of the private sector, U.S. government purchasing can play a role here.
- *Invest in better software, hardware, and infrastructure:* The Intelligence Community should gain a clear understanding of what software and hardware it is relying upon, and then prioritize efforts to transition to better software, hardware, and infrastructure security, and advocate within the U.S. government for measures to improve security broadly in both the government and private sectors.
- *Put serious effort into understanding IoT metadata and its utility:* In this scenario, innumerable aspects of daily existence are affected by the IoT. These devices already spew a remarkable torrent of metadata (Internet traffic patterns, location information). It is the committee's view that this trend is likely to continue, and with a vast increase in IoT devices, could prove a fountain of unexpected information. This will surely be a long-term effort as new forms of IoT are constantly being introduced to the market.
- *Understand the vulnerabilities of 5G/6G, particularly its software-defined aspects, and how they might be exploited:* 5G and the nascent 6G technologies bring a number of changes to mobile telephony infrastructure, most notably in how services are deployed across distributed clouds. This is new and highly complex technology for most telecommunications companies, and it is unclear how well it will be implemented and operated; on the face of it, there may be substantial chances for new vulnerabilities.
- *Continue to grow "star" IT talent within the Intelligence Community:* Of course, the Intelligence Community will always need expert technologists to build and operate systems, but in this scenario highly talented engineers and technicians will become more and more valuable to the Intelligence Community. Such talent is obviously in very high commercial demand, and as such commands reimbursement far beyond what the U.S. government can pay. Nonetheless, the Intelligence Community has some of these "stars" for whom

mission and patriotism outweigh financial gain. In this scenario, the Intelligence Community should do everything within its power to recruit, hire and retain these stars.

SCENARIO 6: COLONY COLLAPSE

Scenario 6, "Colony Collapse," the final scenario the committee chose to explore for this report, explores a notional future described in Table 5.6. A variation on Scenario 6 is offered in Box 5.2.

In 2024, a European mathematician published a paper about a new more efficient way of factoring large numbers. Other researchers noticed the paper and began to explore its nuances, expand on its theoretical basis, and develop experimental software to assess its potential impact. Initial results were ambiguous, but some researchers continued to pursue the direction. Two years later, in 2026, a team of researchers from Europe, Japan, and the United States announced the ability to factor 1024-bit keys for the RSA encryption algorithm much faster than any previous result. More importantly, their technique would enable factoring of the 2048-bit keys in common use in only a million core-years, a level that would be feasible for governments or large companies or criminal groups.

Vendors and standards bodies responded with disbelief and confusion. Some vendors proposed removing support for 2048-bit keys from their products—in 10 years. Standards bodies argued about the best way to adapt to the new threat. The academic research community was confident that this advance would not affect the classical security of elliptic curve cryptography, but these algorithmic details typically got lost in translation in the media. Commercial products that incorporated post-quantum encryption algorithms which would not be affected by rapid factoring were still years in the future, and most current software could not accommodate RSA keys longer than 2048 bits, much less the key lengths of post-quantum algorithms. Some vendors suggested that users change 1024-bit keys weekly or move to discrete logarithm-based algorithms like Diffie-Hellman, even though the algorithmic advance was, after a bit of confusion, extended to this case as well within a few years. When asked, NSA referred reporters to its Commercial National Security Algorithm Suite defined in the Committee on National Security Systems Policy 15 (CNSSP-15), which permits use of RSA and Diffie-Hellman with 3072-bit or larger keys.

TABLE 5.6 Scenario 6: Colony Collapse

Scenario Title	Driver Endpoints	Highlight
Colony Collapse	Disruptive Fragmented Chaotic	A breakthrough in factoring and a lack of focus on systems and security puts information at risk. Despite advances in computing on encrypted data, trust remains low.
Scenario Description		
This scenario posits a breakthrough in the form of a new classical factoring algorithm. Such a breakthrough would render the public key encryption algorithms we rely on more easily attacked with much less effort than today, including by conventional computers. In this case, there would be much less need for a quantum computer. A roomful of powerful servers might be sufficient. Compared to a quantum computing breakthrough, a factoring breakthrough would probably have less advance notice, be easier to keep secret, and be attainable by more countries. In addition, a lack of focus on systems and security puts information at risk in this scenario. Despite advances in computing on encrypted data, trust remains low. This suggests, overall, a much more chaotic world combined with, as in the other scenarios, a more fragmented world politically. In this scenario, the Intelligence Community has to be on the lookout for a breakthrough that will be less easily discerned and more easily hidden than a large-scale quantum computer. Such a breakthrough could come via the efforts of a government or from academic cryptographic researchers. In either case, there might not be advance notice that such a breakthrough was about to occur or had occurred. Experts will debate the likelihood of such a factoring breakthrough, but it seems prudent to ask what would happen should it occur and what steps would need to be taken at that point. As in the other scenarios, much would depend on the ability of the U.S. government to mobilize the private sector to take the necessary steps as well. As with the other scenarios, international fragmentation—of technical standards as well as governmental relations—complicates the challenges for the Intelligence Community. Many governments react to internal tensions and fragmentation by weakening or imposing limits on the use of encryption, further facilitating attacks on encryption. In some respects, the results in this scenario are potentially more widespread and more dangerous than in the other two scenarios, giving added urgency to examinations of the likelihood and consequences of this breakthrough happening.		

> **BOX 5.2**
> **A Variation on Scenario 6 (Colony Collapse): Other Disruptions Can Have Significant Impacts**
>
> Breakthroughs in computation on encrypted data allow social media companies to offer "privacy-preserving" services that sustain their advertising businesses while claiming to protect customers' personal information. Some government and commercial database operators also take advantage of these breakthroughs to implement secure processing of government (e.g., tax) and commercial (e.g., medical insurance) records. Advanced cryptography has become an advertising buzzword: larger, well-resourced companies are able to develop custom protocols implementing these privacy-preserving properties, but other vendors simply make security claims for their products without vetted implementations, and average consumers have no way of telling the difference.
>
> Many countries keep highly detailed information on all citizens, based on tracking their movements, monitoring their communications and purchases, and collecting metadata from smart devices. Some use privacy-preserving technologies but most do not, and such lightly protected databases are easy pickings for intelligence agencies. Commercial companies also keep such detailed profiles, but generally protect the data somewhat better than most governments. Even for the best systems, however, buggy implementations lead to a steady drip of data breaches and disclosures that degrade users' trust. Intelligence agencies' access to these privacy-preserving databases depends on luck and circumstances and is not reliable.
>
> The United Kingdom and Australia mandate government access to encrypted information (personal devices, emails, text messages) in the interest of law enforcement and national security and require service providers or manufacturers to decrypt on demand. One vendor pulled out of Australia and the United Kingdom, but others comply. China mandates the use of national standard encryption algorithms that are presumed breakable by the Chinese government but makes claims for citizens' trust because it does not require vendors to support law enforcement access. All device vendors integrate the Chinese algorithms into their products. Discoveries of bugs in the United Kingdom and Australian exceptional access protocol and users' suspicions lead to a reduction in online activity. One or more incidents occur in which authorized officials (either vendor or government) operate side businesses selling users' keys and/or information.
>
> With changes of government policy worldwide, the deep sharing associated with the Five Eyes arrangement becomes more and more transactional, while sharing of specific intelligence with a broader array of countries becomes a reality. Like the United Kingdom in the 1920s and 1930s, the United States is required to step up its collection against both partners and adversaries. The expanded signals intelligence workload runs head on into citizens' skepticism of government. It is hard to hire the talented and trustworthy people needed to collect and analyze intelligence against a growing list of targets. Weak cryptographic systems and cryptanalytic breakthroughs make the task easier but diversity and sheer numbers make it harder.

While vendors and standards bodies disputed a path forward, hackers, cybercriminals, and nation-states were quick to exploit the new breakthrough. Malware was found circulating with a forged digital signature from a trusted 1024-bit key in an obscure part of Microsoft's code-signing infrastructure related to extended long-term support of legacy operating systems. This was the beginning of a deluge of forged signatures. Forged software updates were a huge problem: a number of automobiles and wireless devices were rendered completely inoperable and had to be replaced. Worse yet, malicious updates to a wide range of devices began to cause havoc: home video cameras began streaming to the public at large, microwave ovens overheated, one type of gasoline pump kept pumping gas even after the tank was overflowing.

As vendors began a "rapid transition" to post-quantum encryption, they encountered further obstacles. NIST and agencies of several other governments had certified multiple post-quantum public key signature and encryption algorithms by 2023 but the certified algorithms were incompatible with each other, and the protocol specifications and software that would be needed to use those algorithms were late (2032–2033) and buggy. In an abundance of caution, NIST had specified that implementations of its standardized post quantum algorithms be used in combination with classical public key algorithms, so that both would have to be defeated to accomplish a successful attack. This too caused issues, as the resulting complexity led to bugs in many implementations. Some industry-specific standards (e.g., wireless, automotive) simply ignored NIST's guidance in favor of specialized algorithms whose security against attack had not been thoroughly vetted.

The multiplicity of standards and the residue of a global Internet and global commerce mean that most products and systems must implement several PQC suites (at least NIST, China, one other). The complexity of the resulting protocols and software leads to implementation errors that introduce further points of attack. A few "minimal" implementations for governments or security enthusiasts appear to be sound but achieve little or no adoption because of their (lack of) usability and functionality.

Warning Signals for Scenario 6

Table 5.7 summarizes some key developments associated with this scenario along with warning signals that might indicate to the Intelligence Community that this scenario is emerging.

Risks and Opportunities in Scenario 6

Scenario 6, "Colony Collapse," is highly complex and dysfunctional. Great powers are jockeying for territory and influence across a series of constantly changing alliances, while all advanced nations suffer near daily breakdowns in some aspects of their appliances, services, and IT infrastructure. The combination of Disruptive, Fragmented, and Chaotic will prove highly challenging for the Intelligence Community: not only will it be extremely hard to predict or shape events, but even routine actions such as holding a virtual meeting may be disrupted without much warning. Employee morale may be very poor given such a difficult work environment.

The lack of mature hardware/software systems has a profound impact on this scenario. If software were better prepared and controlled, it might be a manageable task to determine which systems are susceptible to malfeasance and avoid them. In this scenario, however, everything from databases to doorbells has software of unknown and fluid provenance. This would seem likely to engender a high degree of distrust in all such systems, as one never knows what will fail next.

Risks and opportunities for the Intelligence Community have some similarities with the Scenario 1, "Quantum Breakthrough," but differ in two key points:

TABLE 5.7 Key Developments and Warning Signals for Scenario 6: Colony Collapse

Development	Warning Signal
"In 2024, a European mathematician published a paper about a new more efficient way of factoring large numbers. Other researchers noticed the paper and began to explore its nuances, expand on its theoretical basis, and develop experimental software to assess its potential impact. Initial results were ambiguous, but some researchers continued to pursue the direction. Two years later, in 2026, a team of researchers from Europe, Japan, and the United States announced the ability to factor 1024-bit keys for the RSA encryption algorithm much faster than any previous result."	Has the Intelligence Community seen improvements in number field sieve or other algorithms that may open up new research directions related to factoring, discrete log, or other problems underpinning major public key standards?
"Software updates were a huge problem: a number of automobiles and wireless devices were rendered completely inoperable and had to be replaced. Worse yet, malicious updates to a wide range of devices began to cause havoc: home video cameras began streaming to the public at large, microwave ovens overheated, one type of gasoline pump kept pumping gas even after the tank was overflowing."	Is there a marked increase in the prevalence of malware or bogus software updates with apparently valid digital signatures?

- It is more difficult to see a classical analytic breakthrough coming than a full-scale quantum computer, so the timelines to react are much more compressed.
- Lack of mature hardware/software systems leads to a much higher degree of chaos.

The sections below first list those risks that are shared with the "Quantum Breakthrough" scenario, so the reader does not need to flip pages back and forth, then explicitly call out additional risks for this scenario. These risks are *shared*:

- *Recruiting, hiring, and retaining the "best and the brightest":* Employee quality is a key risk. Will the United States and the Intelligence Community have the talent to be successful, particularly if the United States ceases to attract students and faculty from around the world?
- *Weakening and shifting alliances:* In a fragmented scenario, the United States should not assume that traditional long-term allies remain. As trust and collaboration recede, collection efforts grow more expensive and will surely need deep triage. With ever-shifting bilateral alliances, it may be hard (slow) to pivot to new targets or allies; and it is unwise to share important secrets with any other nation which seems likely to switch alliances.
- *Too many targets:* If each country, or each region, is using its own unique system or form of encryption, it is possible that the Intelligence Community would not have the resources to monitor and penetrate all systems of interest. Fragmentation of the scenario's information infrastructures increases the cost of intelligence collection from information systems, as every system requires its own tailored implementation.

Risks for the Intelligence Community in this scenario that are *not shared* with the "Quantum Breakthrough" scenario include the following:

- *Analytic breakthroughs could create very high risk at short notice:* Encrypted communications or stored data whose protection relied on vulnerable public-key encryption that an adversary has collected or saved may—without much warning—suddenly be at risk of decryption and exploitation. Once such information is collected, there is nothing the originator can do to protect it, but there is a need for an inventory, assessment of potential damage, and development of mitigation plans. In addition, any stolen data that was secured at rest by such encryption would also be revealed.
- *Reliance on untrustworthy hardware/software systems could create very high risk:* As employees increasingly rely on everyday systems of uncertain quality, there is a real chance that cryptanalytic breakthroughs could magnify the risk of collapse of "everyday" appliances and services that their operators had presumed were protected from malicious access by strong encryption. In the worst cases, such breakthroughs could make it extremely difficult for employees to perform their jobs. Of course, this risk also carries over into official government systems that employ commercial hardware or software.
- *Perennial need for high-quality mathematicians:* In order to maintain technical expertise, the Intelligence Community requires a steady stream of students to receive a high-quality, Ph.D.-level education in mathematics or computer science. Those students should be U.S. citizens able to be approved for a security clearance. Historically, the U.S. government has hired mathematicians without background in cryptography and trained them itself. This approach has worked well in the past but appears to be falling behind. In particular, the transition to PQC requires expertise in different areas of mathematics from classical public-key cryptography. In 2015, researchers from Government Communications Headquarters (GCHQ), a key Five Eyes partner to NSA, took the unusual step of engaging with the open cryptographic research

community about the cryptanalysis of a candidate post-quantum lattice-based cryptographic scheme they had designed.[5,6,7,8]

Opportunities for the Intelligence Community in this scenario include the following:

- *Easy pickings:* A "Fragmented and Chaotic" scenario presents a wealth of attack/offensive opportunities for the Intelligence Community. There will be many divergent implementations, each with its own bugs to be exploited; and many suddenly vulnerable to cryptanalytic attack (decryption, use of malign signatures, etc.). The lack of interoperability owing to regionalization might also lead to more weakly encrypted information being accessible. In addition to more opportunities, the attacks can be more aggressive because regionalization will limit the potential for reuse of attack techniques on U.S. or allied systems.
- *Misattribution of attacks:* Fragmented society and governance plus chaotic systems might enable the U.S. government to hide an offensive cryptanalytic breakthrough if they discover it and disguise the resulting exploitation as software exploitation.
- *Spurred investment in cryptographic defense:* An actual cryptanalytic advance would certainly be a great opportunity to push for more investment in defense. It would be well after the fact, but there's nothing like a crisis to spur development.

Actions Relevant to Scenario 6

The sections below first list those actions that are shared with the "A Brave and Expensive New World" scenario, so the reader does not need to flip pages back and forth, then explicitly call out additional risks for this scenario. These actions are shared:

- *Move to mature systems:* This will require enormous effort across the United States, as the current status is at the "chaotic" end of the spectrum for nearly all systems. It is not clear that anyone really understands the implications of good cybersecurity because there is no real-world experience to draw on. It is also unclear how to move toward such systems in practice although the 2021 cybersecurity executive order appears to be an attempt to move in that direction. This issue is much larger than the Intelligence Community and would require effort from major companies, researchers and start-ups, educational institutions, and the U.S. government. It is likely that very heavy investments would be required for research but especially for real-world implementations. As mature systems become available, the Intelligence Community must introduce new, more resilient IT very quickly (not over decades) and as needed abandon less-important IT infrastructure that cannot be quickly secured.
- *Focus on ensuring that the United States can provide an adequate supply of high-quality, trustworthy staff:* Although the large number of foreign-born students in STEM is a major blessing for our country, it should not be taken as a given and may not continue. The U.S. government should work toward considerable improvement in K–12 mathematics and science education, in particular. Advanced cryptographic research is poorly and unevenly funded in the United States; steady, long-term funding should be available for graduate students in cryptography and for efforts that may lead to mature systems (e.g., practical assured programming languages).

---

[5] P. Campbell, M. Groves, and D. Shepherd, n.d., "Soliloquy: A Cautionary Tale," https://docbox.etsi.org/Workshop/2014/201410_CRYPTO/S07_Systems_and_Attacks/S07_Groves.pdf.

[6] Cryptanalytic Algorithms Group, "Soliloquy," Google Groups Conversation, original post February 8, 2015, https://groups.google.com/g/cryptanalytic-algorithms/c/GdVfp5Kbdb8.

[7] P. Campbell, M. Groves, and D. Shepherd, n.d., "Soliloquy: A Cautionary Tale," https://docbox.etsi.org/Workshop/2014/201410_CRYPTO/S07_Systems_and_Attacks/S07_Groves_Annex.pdf.

[8] R. Cramer, L. Ducas, C. Peikert, and O. Regev, 2016, "Recovering Short Generators of Principal Ideals in Cyclotomic Rings," February 25, https://web.eecs.umich.edu/~cpeikert/pubs/logunit.pdf.

- *Limit/mitigate fragmentation, both in technologies and between the United States and allies:* The Intelligence Community and the United States will need to consult with current allies, and try to leverage purchasing power to push common standards and increase implementation maturity. The Intelligence Community will need to emphasize its efforts to preserve existing relationships with allies, but plan to potentially pivot toward new, perhaps short-term relationships. Such a pivot will require both political resources and tending such relationships. Last, because technology fragmentation is a key feature of this scenario, the Intelligence Community needs to learn new technologies and standards; there will be many more of them than at present, and with a higher rate of disruptive scientific advances.

An action for the Intelligence Community that is unique to this and *not shared* with the "Quantum Breakthrough" scenario:

- *Continuously analyze the potential real-world impacts of a cryptanalytic breakthrough:* This should be done on an ongoing basis because reliance on cryptographically secured systems continues to grow rapidly. Efforts to protect U.S. critical infrastructure would be a good starting point: for example, how would a sudden breakthrough affect the U.S. financial system, electric grid, food distribution, hospitals, and so on. This could be augmented with analysis of the impact to employees' daily lives for crippling of such systems, or other "everyday" technology such as refrigerators, furnaces, automobiles. As the scope of such potential vulnerabilities becomes clearer, perhaps some "Plan B" or remediation approaches could be thought out.

SUMMARY OF RISKS AND OPPORTUNITIES

Although the exploration of these three scenarios allowed committee members to explore a variety of different possibilities, one clear objective of this effort was to identify those areas that would be common across the scenarios. Preparing for the risks and opportunities that appear in multiple scenarios will be one way that the Intelligence Community can increase the likelihood that it will be able to mitigate those risks or take advantage of those opportunities as they begin to emerge. Table 5.8 provides a summary of common areas across the scenarios.

SUMMARY OF ACTIONS

In each of the scenarios described above, the committee identified specific actions the Intelligence Community could take to prepare for the risk and opportunities that might present themselves. Table 5.9 highlights those actions that are relevant in two or three of those worlds. These actions would represent the "no regret bets" for the Intelligence Community, because they would provide a benefit, regardless of the operating environment.

TABLE 5.8 Commonalities Across the Scenarios

Risk	Description
Weakening or shifting alliances • Scenario 2 • Scenario 5 • Scenario 6	In all three scenarios, the Intelligence Community is faced with the challenge of building alliances. While each of these scenarios includes the "Fragmented" Society and Governance driver, that is no accident, because the committee believes that it is the most relevant to explore. For this risk, it appears that the endpoints in the other two drivers are not irrelevant, but simply provide different details to a similar narrative: The Intelligence Community, and the United States, will need to invest resources into strengthening existing alliances and/or building new ones over the coming decades. With ever-shifting bilateral alliances it may be hard (slow) to pivot to new targets or allies; and it is unwise to share important secrets with any other nation. The Intelligence Community, and the United States, will need to consider how it shares information and perhaps look more to transactional, ad hoc, arrangements, rather than traditional alliances that dominated the previous decades.
Recruiting, hiring, and retaining the "best and brightest employees" • Scenario 2 • Scenario 6	In both Scenarios 2 and 6, finding and retaining qualified personnel becomes a challenge. Both are at the "Fragmented" and "Disruptive" endpoints of their respective driver, and this combination will present a challenge. A few reasons for this challenge emerged from discussions among committee members: • The U.S. education system is not generating the kind of talent today that will be needed to meet the demand of Scenarios 2 and 6. Much of the talent is found overseas and hiring, clearance, or immigration restrictions could pose a challenge. • In these scenarios, the Intelligence Community is not necessarily the "employer of choice" for tackling large encryption challenges anymore. Anecdotally, the committee understands that many talented cryptographers are joining startup commercial blockchain companies. The Intelligence Community is likely to face increasing competition from corporate and international employers and new "hot" technologies such as cryptocurrencies.
Multitude of targets • Scenario 2 • Scenario 6	In both Scenarios 2 and 6, the combination of "Fragmented" and "Disruptive" seems to drive each country to seek its own unique system or form of encryption. If that is the case, then it is possible that the Intelligence Community would not have the resources to monitor and penetrate all systems of interest. With less data crossing borders, and higher "walls" protecting the information, the Intelligence Community will not be able to focus resources on choke points and will have to rely on more resources or identify "insiders" within the countries themselves to support data collection.

TABLE 5.9 Summary of Preparations That Would Be Beneficial in Two or More Worlds

Action	Description
Move to mature systems • Scenario 2 • Scenario 6	This will require enormous effort across the United States as the current status is at the "chaotic" end of the spectrum for nearly all systems. It is not clear that anyone really understands the implications of good cybersecurity because there is no real-world experience to draw on. It is also unclear how to move toward such systems in practice, although the 2021 cybersecurity executive order appears to be an attempt to make an initial move in that direction. This issue is much larger than the Intelligence Community and would require effort from major companies, researchers and start-ups, and the U.S. government. It is likely that very heavy investments would be required for research and especially for real-world implementations.
Focus on ensuring that the United States can provide an adequate supply of high-quality, trustworthy staff • Scenario 2 • Scenario 5 • Scenario 6	Although the large number of foreign-born students in STEM is a major advantage for the United States, it should not be taken as a given and may not continue. The U.S. government should work toward considerable improvement in K–12 mathematics and science education in particular. It should also encourage college and university computer science and software engineering departments to treat competence in building secure systems as a fundamental requirement for all students. Advanced cryptographic research is poorly and unevenly funded in the United States; steady, long-term funding should be available for graduate students in cryptography and for efforts that may lead to mature systems (e.g., practical assured programming languages).
Limit/mitigate fragmentation, both in technologies and between the United States and allies • Scenario 2 • Scenario 6	The Intelligence Community and the United States will need to consult with current allies, mobilize other government agencies including NIST, and try to leverage purchasing power to push common standards and increase implementation maturity. The Intelligence Community will need to emphasize its efforts to preserve existing relationships with allies, but plan to potentially pivot toward new, perhaps short-term relationships. Such a pivot will require both political resources and tending such relationships. Last, because technology fragmentation is a key feature of each scenario, the Intelligence Community needs to learn new technologies and standards; there will be many more of them than at present, and with a higher rate of disruptive scientific advances.
Focus on alliances • Scenario 2 • Scenario 5 • Scenario 5	While the alliances might look different in 2040 (and in each scenario), the Intelligence Community will still depend on partners to support its offensive and defensive operations. Whether those alliances are long-standing and deep-seated or ad hoc and built to meet specific needs will be determined by the operating environment. The Intelligence Community will need to take steps today, however, to prepare for a more fragmented world in 2040.

6

Implications for U.S. Intelligence

The committee believes that concern regarding the future of encryption that triggered the Office of the Director of National Intelligence (ODNI) to request this study is well founded. The future raises significant, if not critical, issues for the Intelligence Community. Regardless of which scenario envisioned by this study (or another one not so envisioned) develops, the committee believes that encryption will change in fundamental ways that will pose challenges across all aspects of intelligence. Simply put, in an increasingly digital world there will be much more data, more and more of which will be encrypted. At the same time, efforts to attack or disrupt encrypted systems—government, personal, and private sector—will also increase. In short, the committee is positing that encryption will be a far more significant factor in the Intelligence Community's discharge of its mission—even without regard to specific developments in quantum computing.

Encryption is both a necessary capability for U.S. intelligence in protecting various intelligence activities and communications and also a challenge when it comes to collecting intelligence. Offense and defense, or decryption and encryption, are both absolute necessities for successful intelligence operations. This duality also complicates how the Intelligence Community can or should deal with future encryption issues. At a minimum, the future of encryption is an issue for collection, analysis, and security, but more broadly, for the recruiting, retention, operations, and political and intergovernmental relations functions of the Intelligence Community. Most significantly, given the unpredictability of many of the trends discussed in this report, as previously noted, there will be a premium on accurate detection of trends at the earliest possible stage and managing the risks of incorrect assessment. In short, the future will be different in several important ways for the Intelligence Community.

THE DIFFERENT ROLE OF ENCRYPTION IN THE PAST

For a variety of reasons, encryption played a more circumscribed role in the past. One reason is that the United States, and U.S. intelligence, have long been leaders in encryption and related areas, relative to foreign adversaries. As a result, in the absence of significantly increasing technical challenges that threatened to thwart its mission, the Intelligence Community had the luxury of continuing its relative superiority in this area. But now, with more adversary nations (especially China) seeking and making advances in encryption and as academic researchers (especially in Europe) continue to invest in cryptography and advance the theory and practice of encryption, especially of applied cryptography, the committee believes that the advantage that the Intelligence Community enjoyed in this area will diminish if not disappear.

FINDING 6.1: With more adversary nations (especially China) seeking and making advances in encryption and as academic researchers (especially in Europe) continue to invest in cryptography and advance the theory and practice of encryption, the advantage that the Intelligence Community enjoyed in this area will diminish if not disappear.

Another factor is the sheer increase in the number of adversaries to which the Intelligence Community must pay attention. Since 1991, U.S. national security has not had the luxury of dealing primarily with one dominant threat, as was the case during the Cold War. Multiple concerns demand equal attention as the most compelling or potentially most threatening. As DNI James Clapper noted in his Worldwide Threat Assessments in 2012 and 2013, "it is virtually impossible to rank—in terms of long term importance—the numerous, potential threats to U.S. national security ... it is the multiplicity & interconnectedness of potential threats ... that constitute our biggest challenge."

Recognition of the variety of threats was a major factor in the creation and adoption of the National Intelligence Priorities Framework (NIPF) in 2003. The NIPF is driven by a prioritized list of issues, as determined by the members of the National Security Council. The committee assumes that there is no NIPF issue for the "future of encryption" per se. This issue likely falls within two broader issues that have also been noted by DNI Avril Haines in her 2021 Annual Threat Assessment of the U.S. Intelligence Community: cyber and emerging technology. The committee further assumes that cyber is already in the top priority tier and that emerging Technology is either in that top tier or one close to it.

Thus, with a greater number of adversaries generally, a wider range of threats to national security, increasing uses of encryption in all manner of communication and information storage, and ever more sophisticated adversaries using encryption (for both offense and defense), the Intelligence Community will need to deal with the issue of encryption in a more direct and profound way. As explained below, this raises broad questions of whether the Intelligence Community is most effectively positioned to undertake that task, and how the substantive activities of the Intelligence Community might have to adapt to a future world in which encryption is a more problematic challenge.

EFFECTIVE STRUCTURES TO ADDRESS ENCRYPTION

An initial question for Intelligence Community leaders is how best to handle the issue of the future of encryption. At a high level, the committee also notes that although policy makers may appreciate in broad terms that encryption is an important, serious, and complex issue, few of them will have the technical expertise to understand the issue in any depth. Therefore, it is incumbent upon the Intelligence Community to make clear to policy makers what is at stake, what is the nature of the near-term and longer-term threat, what are the opportunities, and what are U.S. options to deal successfully with the full array of encryption issues. On an operational level, major responsibilities will likely fall to the National Security Agency and also U.S. Cyber Command, but neither of these is an all-source analytic agency. Fully addressing encryption topics would therefore require participation by analysts from the Central Intelligence Agency and the Defense Intelligence Agency (DIA), and perhaps the Department of State's Bureau of Intelligence and Research (INR), which are the three all-source analytic agencies.

Three further issues arise. The first issue is whether or not there currently exists within the Intelligence Community the requisite talent and capability to deal with the full range of policy and technical matters that relate to encryption. Past efforts at establishing expertise databases have not been overly successful. If the requisite talent is not on hand, then it will have to be recruited. Here, the Intelligence Community faces the issue of competition with the private sector, which can offer much more lucrative salaries, stock options, and potentially career opportunities. The U.S. government as a whole, and the Intelligence Community in particular may have to adjust personnel policies and expectations to accommodate individuals who do not plan to spend their entire careers in government, or who go in and out of government jobs. This may be especially challenging for intelligence, given security requirements.

An additional directly related matter is the highly dynamic nature of research and developments in the encryption field. For example, although government analysts may be able to monitor academic research in encryption, they may have a much harder time understanding the motivations and plans of commercial entities that integrate encryption into products and ensure the wide proliferation of encryption technologies. It is difficult to keep abreast of these developments, especially for government officials who have full-time jobs away from academia and industry. This further suggests that it may be necessary to refresh the workforce that deals with matters related to encryption on a periodic basis.

Alternatively, the Intelligence Community should consider keeping in touch with, and perhaps contracting with individuals whose knowledge and skills can be beneficial. It is important to keep in mind that security clearances should not be major impediments here. These outside experts are sharing their knowledge and expertise with the Intelligence Community. (Specific situations involving sensitive interests in specific technologies or cases where an association with a person is sensitive can be handled on an individual basis.)

The second issue is the question of how these intelligence efforts regarding encryption across several agencies can be coordinated to ensure the sharing of information about the encryption issue, threats, developments, and opportunities.

The third issue is that the various capabilities discussed in this study—whether to create successful encryption or to defeat encryption—require strong technical capabilities in terms of both technology and people. However, decisions about the use of encryption or efforts to defeat encryption will likely be made by people who do not have and—in general—do not need these same technical skills. These people will be policy makers, who bring a variety of skills and backgrounds to their jobs and to the decisions they make. (As an analogy, consider President Harry Truman's decision to use the atomic bomb. Truman was not a physicist, and he could be given only a rather rudimentary understanding of what the use of the bomb would entail, as none had ever been dropped on a city.) Therefore, it will be imperative that the technological explanations of encryption decisions are made in such a way that non-technical people can understand them and make appropriate decisions. This can sometimes be problematic as there will be a "communications gap" between the technologists and the policy makers. It is incumbent on both groups to try to understand one another but the greater burden will lie with the technologists to ensure that policy makers have a firm and comprehensible basis for their decisions.

It would also be useful for the Intelligence Community to conduct an assessment of current capabilities—collection, analysis, and security—to deal with the encryption issue, as well as a projection of what will be needed in the future.

HOW THE INTELLIGENCE COMMUNITY MAY NEED TO ADAPT TO THE FUTURE WORLD OF ENCRYPTION

The complexities of the future of encryption not only raise organizational issues but also pose challenges to the way the Intelligence Community conducts its mission in practice. Those challenges stem principally from the fact that each of the scenarios described above posits a more fragmented world. A fragmented world suggests a greater multiplicity of hardware and software and perhaps even national Internets that are to some extent separated from the more unified Internet of today. This will increase the range of places and actors that have to be watched and will also likely increase the likelihood of less capable—or more buggy—software. All of this will increase the collection and analysis burden. This can be a threat if some of this software is used within the U.S. government but also an opportunity in terms of gaining access to others' systems if they are using these tools.

A fragmented world also has possible political implications for U.S. intelligence. It can suggest a loosening of alliances or a decline in cooperation among intelligence partners in specific areas, such as signals intelligence or encryption. Democratic governments may come under popular pressure to take on more nationalistic approaches to some issues—as was the case with New Zealand in the 1980s when their anti-nuclear policy led to their "suspension" within the Five Eyes intelligence alliance. Demands for greater privacy or uneasiness about some aspects of international cooperation could be such drivers. The reactions within Europe to the disclosures by Edward Snowden are an example.

Fragmentation involves more than nation states and is a current problem, across both government and private sectors. The United States has gone from a focus on one dominant threat—the Soviet Union—to multiple threats. In military terms, this is still limited to a finite number of nation-states. However, in the realm of cyber and emerging technology, the United States is already dealing with many states and with many non-state actors, either working independently or in league with nation-states. The ubiquity of computer technology and the very low barriers to entry make this multiple threat very real. Although only a few nations may be able to defeat the most sophisticated defenses, experience has shown that many defenses are "chaotic" and subject to attack by numerous unsophisticated actors. It is also possible that the United States may have to deal with hostile non-state "cyber" alliances that will be difficult to discern and to assess. This is a level of multiple threats much greater and more complex than has been the case in the past, again raising issues within U.S. intelligence of responsibilities and coordination.

This report posits that while the availability of a large-scale quantum computer sufficient to attack current public-key encryption is uncertain in a 20-year time frame, other developments in the realms of scientific advances and the integration of encryption into systems are likely to alter the encryption challenges and opportunities facing the U.S. government. The Intelligence Community will be responsible for keeping abreast of these developments, some of which will undoubtedly not occur in the open. This, again, raises issues about collection and analytic capabilities.

Last, the Intelligence Community will continue to be responsible for warning against major cyber intrusions, whether against government systems or in the private sector. The committee sees no reason to expect a decrease in these attacks, which in turn will affect perceptions of and perhaps support for U.S. intelligence.

7

Findings

This chapter consolidates the findings from the previous chapters. For completeness, it also repeats the summaries of risks and actions for the Intelligence Community that were documented at the end of the discussion in Chapter 5 of the scenarios that the committee identified.

FINDINGS FROM CHAPTER 2 (INTRODUCTION TO ENCRYPTION)

FINDING 2.1: Stateful digital signatures based on hash functions are practical today and will remain secure even if large-scale quantum computers are practical or if new number theoretic attacks are developed that affect other quantum-resistant signature algorithms. These algorithms may be appropriate for use in specific scenarios such as firmware signing. While their wide application would pose some difficulties for system implementers, they would provide a viable digital signature option for some use cases in the event that a cryptanalytic breakthrough rendered other digital signature algorithms vulnerable.

FINDING 2.2: For smaller-scale applications within a single sophisticated organization with little tolerance for the possible risks to public-key cryptography (whether from a mathematical advance or quantum computers), it may be possible to use key distribution centers (KDCs) to replace or augment some uses of public-key cryptography. Because of the different trust models and attack surface, deploying KDCs to replace public-key cryptographic functions in open settings like Hypertext Transfer Protocol Secure (HTTPS) would be difficult technically, politically, and logistically.

FINDING 2.3: If an organization has encrypted information in the past using keys negotiated with an algorithm that later becomes vulnerable to cryptanalysis by a quantum or classical computer, there is little that the organization can do at the cryptographic level to prevent future decryption of ciphertexts that have already been intercepted and stored by an adversary. Organizations in that situation may be best served by understanding their risks from decryption of previously encrypted information, assembling an inventory of such information, and taking measures to limit the damage in the event that information is decrypted in the future.

FINDING 2.4: The research community continues to make improvements in the technology of computation on encrypted data. Such improvements can be expected to enable new ways of securely sharing both government and private-sector information.

FINDINGS FROM CHAPTER 4 (DRIVERS)

Findings Pertaining to Scientific Advances

FINDING 4.1: Most of the current public scientific expertise in algorithm design, cryptanalysis, and other areas of applied cryptography is outside the United States, largely in Europe. In contrast, within the United States, cryptography is taught as an area of theoretical computer science. The specific areas of expertise necessary to guide and facilitate the transition to post-quantum cryptography are relatively new and will require a more robust educational pipeline to train new talent.[1] Public research investment, through the National Science Foundation and other organizations, would encourage this process, while strict U.S. export control regulations have historically discouraged talent from locating in the United States.

FINDING 4.2: An improvement in asymmetric cryptanalysis algorithms could have a significant effect on the security of public key encryption algorithms that are in wide use today. Such an improvement would enable more efficient attacks on encrypted information using conventional computers rather than requiring the construction of a quantum computer. Furthermore, it could potentially be exploited in secret and with little or no advance notice.

Findings Pertaining to Society and Governance

FINDING 4.3: It is difficult to predict what mix will occur of low or high levels of government regulation of cryptocurrencies. Low levels of regulation will be subject to criticism for facilitating criminal activity. High levels of regulation will be subject to criticism for excessive surveillance. Market and technological factors further make it difficult to predict future growth in the sector. Of this uncertainty, it is also uncertain the extent to which intelligence agencies will retain, increase, or decrease their access to financial, transactional information.

FINDING 4.4: Forces for both globalization and fragmentation will be present. Even if the committee were in a position to predict whether globalization or fragmentation were more likely to prevail, these trends are complex and interrelated. Some trends reinforce themselves and others prompt opposite reactions. Thus, it is difficult to determine which forces are likely to prevail on any given issue. In theory, this means that the Intelligence Community will need to be prepared for alternative extremes—for example, a world in which authoritarian governments weaken or ban encryption in ordinary communications, and a world in which governments support pervasive use of encryption citing privacy and security concerns. Because that preparation is impossible to sustain over any meaningful period, there will be a premium on accurate detection of trends at the earliest possible stage and managing the risk of an incorrect assessment.

FINDING 4.5: The Internet and increasing technological interdependence promotes globalization. The shared experience of individuals around the globe owing to information and communications being instantly and ubiquitously available is a powerful cause of international commonality. That factor, along with convergence of technologies, ever-increasing global interdependence on all levels and across economic and political sectors, the continued growth of world trade and the likely ongoing increase in the role of the private sector, with

---

[1] To understand the cryptographic landscape, one must receive a Ph.D. in cryptography with at least 3–5 years of highly specialized training in graduate school. Even though the information is freely available on the Internet, the sheer volume of information and high degree of specialization means that without hands-on advising, it is nearly impossible to learn the skillset necessary to become proficient in cryptography.

its constant drive for efficiency and common standards, will tend to powerfully mold the world in a unified way, increasing the likelihood that nations around the world will take similar approaches to issues relevant to encryption.

FINDING 4.6: Governmental regulation, for better or worse, of communications technology may lead to fragmentation on national lines. National security concerns have the effect, whether specifically intended or not, of creating competing national technologies—by limiting the exports of sensitive technology or by curtailing imports of equipment that may permit surreptitious surveillance by a foreign manufacturer or its government. Potent forces are present, for both beneficial and malicious reasons, that could predispose the global arrangement toward individual nationalistic or regional solutions to issues bearing on encryption. In many countries, there is growing support for "digital sovereignty," a term that can mean various things ranging from having regulatory decisions made nationally instead of by Silicon Valley, and support for protectionist trade policies, to segmenting the Internet by blocking communications with other countries. In addition, national regulations to promote online competition, enhance cybersecurity, curtail hate speech, and protect citizens' data privacy might well vary significantly around the globe and even in geopolitical regions where there might otherwise be commonality. A rise in citizens' mistrust of governments (especially in the area of surveillance) might lead to a corresponding growth in the use of encrypted communications (both to avoid government surveillance and in response to general privacy concerns). Moreover, individual countries or blocs of like-minded countries might impose (or continue to impose) substantive communications content requirements enabled by technological distinctions at national levels, including, for example, banning or discouraging end-to-end encryption (so as to permit government surveillance), or mandating a variety of governmental access to otherwise encrypted communications (perhaps through required turnover of encryption keys to authorities or insisting on the use of specified encryption schemes).

Findings Pertaining to Systems

FINDING 4.7: In most cases, a common set of security protocols and cryptographic algorithms are used globally, and systems and networks today are largely interoperable. This may not remain the case; the factors that led to this interoperability are weakening, and pressures to create national and regional differences are growing.

FINDING 4.8: In every scenario, bugs in software and operational errors are the weakest links in security.[2]

FINDING 4.9: Communications and storage depend on a software stack: hypervisor (a program that allows a computer to run several operating systems simultaneously), operating system, libraries, and application. While quantum computers or mathematical advances are important research topics, bugs or operational mistakes in this stack are the biggest source of system insecurity. Exploiting these errors is, and likely will remain, the biggest opportunity for offense, and minimizing them the highest priority for defense.

FINDING 4.10: The United States needs far more data security expertise than is currently available, and these needs are growing substantially. The failure to meet these needs could have significant and widespread ramifications both for national security and the private sector. All software developers and computer scientists require basic competence in computer security. In addition, a growing number of people will require deep expertise in security. The required skills are not easy to teach, as students need both security-focused knowledge and a deep technical knowledge across multiple subjects and layers of abstraction. If the U.S. educational system does not meet these needs, or if the United States becomes a less attractive destination for students, researchers, and entrepreneurs born in other countries, the shortage will be much worse. Technological changes

---

[2] T. Armerding, 2016, "The OPM Breach Report: A Long Time Coming," CSO Online, October 13, https://www.csoonline.com/article/3130682/the-opm-breach-report-a-long-time-coming.html.

may rapidly increase demand for rare skills or may reduce demand by enabling tasks that currently require exceptionally skilled individuals to be performed by a broader range of people.

FINDING 4.11: Practical knowledge about the security of cryptographic systems will continue to be widely disseminated across the globe. Effective work (offensive or defensive) can be performed by a few skilled individuals. As a result, unlike areas where a country can obtain dominant capabilities by incurring costs that other countries cannot afford, many countries will have significant data security capabilities and none will be dominant.

FINDING 4.12: The transition to post-quantum cryptography is likely to be prolonged over many years. It may also provide a rationale for replacing obsolete systems that have other security problems.

FINDING 4.13: The complexity of the transition to post-quantum cryptography will likely introduce a range of new security vulnerabilities.

FINDING 4.14: A new classical cryptanalysis algorithm or quantum computing development could result in rushed and disorganized efforts to replace widely used public key algorithms or other cryptographic standards. Such a breakthrough would require mitigation efforts that would be more complex than fixing typical software bugs, such as the coordinated deployment of major protocol updates across implementations and services.

FINDING 4.15: 5G may introduce a number of new systems issues in practice, owing to both complex new suites of software and operator inexperience in distributed cloud environments.

FINDING 4.16: Many Internet of Things (IoT) components are poorly secured and easy to subvert, with an extremely wide range of consequences that are difficult to predict but potentially very high impact for the Intelligence Community and broader society. Because IoT will likely bring significant improvements to many aspects of life, however, more money and energy may be devoted to securing such devices going forward.

FINDING FROM CHAPTER 6 (IMPLICATIONS FOR U.S. INTELLIGENCE)

FINDING 6.1: With more adversary nations (especially China) seeking and making advances in encryption and as academic researchers (especially in Europe) continue to invest in cryptography and advance the theory and practice of encryption, the advantage that the Intelligence Community enjoyed in this area will diminish if not disappear.

Table 7.1 presents a summary of the risks and opportunities that would be realized if the various scenarios were to come to pass. Table 7.2 presents actions that could be of benefit if the scenarios were to occur.

TABLE 7.1 Summary of Risks and Opportunities from Scenarios

Risk	Opportunity
Weakening or shifting alliances • Scenario 2 • Scenario 5 • Scenario 6	In all three scenarios, the Intelligence Community is faced with the challenge of building alliances. While each of these scenarios includes the "Fragmented" Society and Governance driver, that is no accident, because the committee believes that it is the most relevant to explore. For this risk, it appears that the endpoints in the other two drivers are not irrelevant, but simply provide different details to a similar narrative: The Intelligence Community, and the United States, will need to invest resources into strengthening existing alliances and/or building new ones over the coming decades. With ever-shifting bilateral alliances it may be hard (slow) to pivot to new targets or allies; and it is unwise to share important secrets with any other nation. The Intelligence Community, and the United States, will need to consider how it shares information and perhaps look more to transactional, ad hoc, arrangements, rather than traditional alliances that dominated the previous decades.
Recruiting, hiring, and retaining the "best and brightest employees" • Scenario 2 • Scenario 6	In both Scenarios 2 and 6, finding and retaining qualified personnel becomes a challenge. Both are at the "Fragmented" and "Disruptive" endpoints of their respective driver, and this combination will present a challenge. A few reasons for this challenge emerged from discussions among committee members: • The U.S. education system is not generating the kind of talent today that will be to meet the demand of Scenarios 2 and 6. Much of the talent is found overseas, and hiring, clearance, or immigration restrictions could pose a challenge. • In these scenarios, the Intelligence Community is not necessarily the "employer of choice" for tackling large encryption challenges anymore. Anecdotally, the committee understands that many talented cryptographers are joining start-up commercial blockchain companies. The Intelligence Community is likely to face increasing competition from corporate and international employers and new "hot" technologies such as cryptocurrencies.
Multitude of targets • Scenario 2 • Scenario 6	In both Scenarios 2 and 6, the combination of "Fragmented" and "Disruptive" seems to drive each country to seek its own unique system or form of encryption. If that is the case, then it is possible that the Intelligence Community would not have the resources to monitor and penetrate all systems of interest. With less data crossing borders, and higher "walls" protecting the information, the Intelligence Community will not be able to focus resources on choke points and will have to rely on more resources or identify "insiders" within the countries themselves to support data collection.

TABLE 7.2 Summary of Actions from Scenarios

Action	Description
Move to mature systems • Scenario 2 • Scenario 6	This will require enormous effort across the United States, as the current status is at the "chaotic" end of the spectrum for nearly all systems. It is not clear that anyone really understands the implications of good cybersecurity because there is no real-world experience to draw on. It is also unclear how to move toward such systems in practice, although the 2021 cybersecurity Executive Order appears to be an attempt to make an initial move in that direction. This issue is much larger than the Intelligence Community and would require effort from major companies, researchers and start-ups, and the U.S. government. It is likely that very heavy investments would be required for research and especially for real-world implementations.
Focus on ensuring that the United States can provide an adequate supply of high-quality, trustworthy staff • Scenario 2 • Scenario 5 • Scenario 6	Although the large number of foreign-born students in STEM is a major blessing for our country, it should not be taken as a given and may not continue. The U.S. government should work toward considerable improvement in K–12 mathematics and science education in particular. Advanced cryptographic research is poorly and unevenly funded in the United States; steady, long-term funding should be available for graduate students in cryptography and for efforts that may lead to mature systems (e.g., practical assured programming languages).
Limit/mitigate fragmentation, both in technologies and between the United States and allies • Scenario 2 • Scenario 6	The Intelligence Community and the United States will need to consult with current allies and try to leverage purchasing power to push common standards and increase implementation maturity. The Intelligence Community will need to emphasize its efforts to preserve existing relationships with allies, but plan to potentially pivot toward new, perhaps short-term relationships. Such a pivot will require both political resources and tending such relationships. Last, because technology fragmentation is a key feature of each scenario, the Intelligence Community needs to learn new technologies and standards; there will be many more of them than at present, and with a higher rate of disruptive scientific advances.
Focus on alliances • Scenario 2 • Scenario 5 • Scenario 5	While the alliances might look different in 2040 (and in each scenario), the Intelligence Community will still depend on partners to support its offensive and defensive operations. Whether those alliances are long-standing and deep-seated or ad hoc and built to meet specific needs will be determined by the operating environment. The Intelligence Community will need to take steps today, however, to prepare for a more fragmented world in 2040.

Appendixes

A

Statement of Task

The National Academies of Sciences, Engineering, and Medicine will convene an ad hoc committee to identify potential scenarios over the next 10 to 20 years for the balance between encryption and decryption (and other data and communications protection and exploitation capabilities). The committee will then assess the national security and intelligence implications of the scenarios it deems most relevant and significant, based on criteria it develops.

The committee will first identify plausible scenarios, and the technological drivers (and other major drivers as deemed relevant by the committee) behind these scenarios, and potential areas of technology surprise. It will consider such factors as likelihood, speed, difficulty of planning and response, and consequence, in order to advise on which scenarios are most worthy of attention. The committee will also consider implications for applications of encryption such as cybersecurity, digital currency, cybercrime, surveillance, and covert communication.

The committee will then assess the national security, intelligence, and broad societal implications of each scenario determined by the committee to be most worthy of attention; identify and assess options for responding to these scenarios; and assess the implications for future Intelligence Community investments. In doing so, it will consider actions common across all scenarios, scenario-dependent actions, and technology developments that the Intelligence Community should monitor in order to narrow the range of possible scenarios in the future. It will also consider how other governments might act in each of the scenarios, and the implications of those actions for the United States. This project will produce a peer-reviewed consensus report.

B

Meeting Agendas

OCTOBER 2, 2020

Held via Teleconference

11:00–11:45	Review of the Last Meeting and Identify Follow-Up Questions and Gaps to Fill
11:45–12:00	Break
12:00–12:50	Scenario-Based Analysis—Hans Davies
12:50–1:30	Discussion and Break
1:30–2:30	Scenario-Based Analysis—Steve Weber
2:30–3:00	Discussion of Possible Scenarios
3:00	Meeting Adjourns

OCTOBER 23, 2020

Held via Teleconference

11:30–12:45	Computing Standards—NIST Computer Security Division
12:45–1:00	Break
1:00–2:00	Quantum Computing: Where Is It Going and When Will It Get There?—Mark Horowitz
2:00–2:15	Break
2:15–3:00	Dual Elliptic Curve Issue—Nadia Heninger

NOVEMBER 6, 2020

11:00–11:30	Closed Session: Review of Progress
11:30–12:30	Open Data-Gathering Session: Topics in Quantum Communication and Encryption—John Manferdelli
12:30–12:45	Break
12:45–1:45	Open Data-Gathering Session: Implementation, Transition, and Standardization During the Move to Quantum-Resistant Encryption Algorithms—Brian LaMacchia and Bob Blakley
1:45–2:00	Break
2:00–3:00	Closed Session: Scenario Exercise for Committee Members
3:00	Meeting Adjourns

NOVEMBER 20, 2020

11:00–11:15	Closed Session
11:15–12:15	Open Session: U.S. and International Law Enforcement Issues—Speaker Darrin Jones
12:15–1:15	Open Session: Encryption, Exceptional Access, and Privacy—Susan Landau
1:15–1:30	Break
1:30–3:00	Closed Session
3:00	Meeting Adjourns

DECEMBER 4, 2020

11:00–11:15	Closed Session
11:15–12:15	Open Session: Directions in Encryption Research—Josh Barron
12:15–12:30	Break
12:30–1:15	Open Session: Public Accountability of Secret Processes—Sunoo Park
1:15–1:30	Break
1:30–2:15	Open Session: Future of Encryption and Cryptography—Dan Boneh
2:15–3:00	Closed Session
3:00	Meeting Adjourns

C

Potential Scenarios

As part of the committee's exploration of the worlds and the down-selection process, the group developed snapshots for each world to discuss how the endpoints would converge. As discussed in Chapter 5 and Chapter 7, the committee determined that Scenario 2, Scenario 5, and Scenario 6 would be the most interesting to explore, given the sponsor's questions and resulting convergence of the endpoints. Table C.1 summarizes all the scenarios that were generated, including those that were not selected for exploration.

TABLE C.1 Short Descriptions of All the Scenarios Generated for Possible Exploration

Scenario	Systems	Scientific Advances	Society and Governance	Description
1	Mature	Predictable	Fragmented	Balance favors defense. No quantum computers and few mathematical breakthroughs; deterioration of international trust leads to splintering of Internet and localized 5G/6G; some use of computation on encrypted data to limit data sharing; Intelligence Community is forced to increase collection efforts to offset lost partnerships; some countries ban end-to-end encryption and keep extensive data on citizens, creating targets for collection; loss of trust in governments amplifies insiders as an offensive opportunity and a defensive threat; growth of untraceable digital currency systems raises challenges to financial intelligence collection; high quality standards and implementation enhance success of IA and make collection challenging.
2	Mature	Disruptive	Fragmented	Balance favors defense. Quantum computers and/or new factoring techniques threaten data previously encrypted with older algorithms; successful post-quantum transition mitigates the damage; deterioration of international trust leads to splintering of Internet and local 5G/6G; wide use of new techniques for computation on encrypted data to limit data sharing; Intelligence Community is forced to increase collection efforts to offset lost partnerships; some countries ban end-to-end encryption and keep extensive data on citizens, creating targets for collection; loss of trust in governments amplifies insiders as an offensive opportunity and a defensive threat; growth of untraceable digital currency systems raises challenges to financial intelligence collection; high quality standards and implementation enhance success of IA and this plus rapid changes in technology make collection more challenging.

continued

TABLE C.1 Continued

Scenario	Systems	Scientific Advances	Society and Governance	Description
3	Mature	Predictable	Global	Balance of offense and defense. No quantum computers and few mathematical breakthroughs; effective security standards and implementations plus limited offensive breakthroughs drive IA mission success; international partnerships and global sharing of commercial/private information help to offset improved technical security; security of 5G/6G, Internet of Things (IoT), and cryptocurrencies benefits from common and sound global standards; common international standards mean that effective offensive techniques often have global reach.
4	Mature	Disruptive	Global	Balance of offense and defense. Quantum computers and/or new factoring techniques threaten data previously encrypted with older algorithms; successful post-quantum transition mitigates the damage; new approaches to computing on encrypted data further challenge offense; international partnerships and global sharing of commercial/private information help to offset improved technical security; security of 5G/6G, IoT, and cryptocurrencies benefits from common and sound global standards; common international standards mean that effective offensive techniques often have global reach.
5	Chaotic	Predictable	Fragmented	Balance favors offense. No quantum computers and few mathematical breakthroughs; poor quality standards and implementations make transition to post-quantum late and bug-ridden; poor implementations lead to opportunities for offense and a nightmare for IA; IoT and 5G/6G suffer from reliance on poor quality code and lack of common standards; Intelligence Community is forced to increase collection efforts to offset lost partnerships; some countries ban end-to-end encryption and keep extensive data on citizens, creating targets for collection; loss of trust in governments amplifies insiders as an offensive opportunity and a defensive threat; fragmented policy environment increases the Intelligence Community workload—every country has its own buggy implementations; pervasive use of poor quality systems facilitates offensive mission.
6	Chaotic	Disruptive	Fragmented	Balance favors offense. Quantum computer breakthroughs and factoring improvements threaten classic cryptography; post-quantum implementations are late and vulnerable; implementations of computation on encrypted data are buggy, leading to easy attacks on the new mechanisms; deterioration of international trust leads to splintering of Internet and local 5G/6G; adoption of computation on encrypted data limited by poor quality standards and implementations; Intelligence Community is forced to increase collection efforts to offset lost partnerships; some countries ban end-to-end encryption and keep extensive data on citizens, creating targets for collection; loss of trust in governments amplifies insiders as an offensive opportunity and a defensive threat; fragmented policy environment increases the Intelligence Community workload—every country has its own buggy implementations; pervasive use of poor quality systems facilitates offensive mission.
7	Chaotic	Predictable	Global	Balance favors offense. No quantum computers and few mathematical breakthroughs; poor quality standards and implementations make transition to post-quantum late and bug-ridden; poor implementations lead to opportunities for offense and a nightmare for IA; despite common standards, IoT, 5G/6G, and cryptocurrencies suffer from reliance on poor quality standards and code; international partnerships and global sharing of commercial/private information amplify the impact of poor technical security; common international standards mean that effective offensive techniques often have global reach.

TABLE C.1 Continued

Scenario	Systems	Scientific Advances	Society and Governance	Description
8	Chaotic	Disruptive	Global	Balance favors offense. Quantum computer breakthroughs and factoring improvements threaten classic cryptography; post-quantum implementations and standards are poor quality, late, and vulnerable; implementations of computation on encrypted data are buggy, leading to easy attacks on the new mechanisms; despite common standards, IoT, 5G/6G, and cryptocurrencies suffer from reliance on poor quality standards and code; international partnerships and global sharing of commercial/private information amplify the impact of poor technical security; common international standards mean that effective offensive techniques often have global reach.

D

Global Trends 2040

The National Intelligence Council's report *Global Trends 2040: A More Contested World*,[1] which was "designed to provide an analytic framework for policymakers early in each administration," was released in March 2021, when the committee was starting to prepare this document. The committee recognized that *Global Trends 2040* reinforces the committee's views about the Society and Governance driver most likely to impact the future of encryption. Accordingly, the committee thought it useful to summarize the most pertinent aspects of *Global Trends 2040*.

Global Trends 2040 focuses on five themes: increasing global challenges, including climate change, disease, financial crises, and technology disruptions; fragmentation of responses by the international community to such challenges, notwithstanding increasing connectivity in all sectors; disequilibrium from the mismatch at all levels between challenges and needs in an international system not suited to address compounding global change; contestation within the international community as tensions, division, and competition increase because of such change; and the existential need for adaptation to these changes. A review of the components of each theme lends support to this report's discussion of the Society and Governance driver.

Global Trends 2040 examines four topics: structural forces, technology, emerging dynamics, and future scenarios, each predicted to impact U.S. national security strategies in the next two decades.

Beginning with *structural forces*, four factors are identified as interacting to alter or disrupt the existing global context:

- *Demographics* lead to a larger aging population in some nations, a general decline in education, health and poverty reduction, and increased migration pressures.
- *Societal changes* result from populations becoming pessimistic and distrustful because of economic, technological, and demographic trends; increasingly siloed information contributes to fault lines supporting civic nationalism, volatility, and a more informed population with the ability to express demands directly.
- *Future scenarios for 2040* include resurgence of open democracies, rapid technological developments, and public–private partnerships that transform the global economy while improving the quality of life, in contrast to efforts by China and Russia to increase societal controls with monitoring techniques.

---

[1] National Intelligence Council, 2021, *Global Trends 2040: A More Contested World*, Office of the Director of National Intelligence, March, https://www.dni.gov/index.php/gt2040-home, accessed October 21, 2021.

- *Environmental changes* increase risks to human life and national security, with an unequal impact with a destabilizing economic impact nationally and internationally.

Meanwhile as the pace and reach of technology increases, global competition challenges local governments, facing a more engaged population. This mismatch between public demand and government capability extends to the international level: no one state controls, although the United States and China continue to have powerful, if apposite, levels of influence.

From the international perspective, without unitary control, the world is adrift and several outcome options compete:

- A directionless, chaotic, volatile world, with rules ignored;
- Competitive co-existence as the norm;
- Separate silos; or
- Tragedy, followed by a more cohesive mobilization and move to common standards.

Second, *technology* reveals increased technological developments that transform human experience and capabilities, but produce tensions and societal and economic disruption among states. Heightened competition for talent, knowledge, and markets results in a multitude of areas (e.g., artificial intelligence, robotics, virtual reality) raising ethical, societal, and security questions about who we are as humans, our environmental impact, and the bounds of warfare. Such technologies also drive further transformation and disruption.

Third, *emerging dynamics* challenges, in the context of societal, state, and international relations, become increasingly interconnected globally in a technologically advanced and diverse world.

- Society becomes increasingly disillusioned, informed, yet divided, as large parts of the global population become insecure, uncertain, and distrustful of all institutions.
- Populations divide and gravitate to more comfortable units, leading to competition in visions, goals, and beliefs.
- Transnational identities, allegiances, and siloed information combine to enhance intra-state fault lines, undermine civic nationalism, and increase volatility.
- Populations gain ability and incentives to demand social and political change, demanding services and recognition; governments are challenged, the international order and traditional norms are contested, democratic governance declines, and the risk of interstate conflict grows.

Fourth, the scenarios *for 2040* posed three questions:

- How severe are the looming global challenges?
- How do states and nonstate actors engage in the world, including focus and type of engagement?
- Last, what do states prioritize for the future?

In response, the report discusses the strengths and weaknesses of the five possible scenarios:

- A renaissance of democracies:
  — Fostering of scientific research and technological innovation catalyzes economies for domestic and international benefit.
  — Better services and anticorruption efforts restore public institutional trust.
  — U.S. leadership is central but requires alliances and international institutions.
  — Repression, stalled economic growth, and demographic pressure may undermine authoritarian regimes in China and Russia so that they become unpredictable, more aggressive neighbors.

- A world adrift:
  — A directionless world without international rules, with limited global cooperation, and with limited technological solutions.
  — An aggressive China increases armed conflict risk with regional powers, especially over critical resources; developing countries are compelled to co-operate.
  — Regional powers and nonstate actors (e.g., corporations) increase influence over cyber, space, and technologies, but without power to dominate.
  — Weakened rules and multilateral cooperation make a more vulnerable world for non-state actors.
  — "Global challenges" increase, as states lack incentives to act collectively and proceed in a "patchwork" approach.
- Competitive coexistence:
  — Global rivals (e.g., the United States and China) compete for markets, resources, and brand recognition under mutually accepted rules, with national corporations agreeing.
  — Economic interdependence reduces armed conflict risk, although influence operations, corporate espionage, and cyberattacks continue.
  — Security relies on managing competition to avoid harming their prosperity or that of the global economy.
  — Climate change remains a challenge for stability.
- Separate silos:
  — Separate economies risk financial loss as supply chains facture, markets are lost, and other sectors decline, even though supply chain disruptions are less problematic.
  — Some large countries and abundantly resourced countries (e.g., the United States and Canada) adapt; others are challenged by self-sufficiency.
  — For domestic stability, states adopt various political models, combining democracy, authoritarianism, surveillance, repression, and exclusionism.
  — Without immigration, innovation lags; more resources go to educating citizens.
  — Focus on climate change, healthcare disparity, and poverty falters; countries adapt and look for risky solutions.
  — Internal security leads to smaller resource conflicts but a focus on scared resources, which diverts attention from domestic problems to support against foreign enemies.
- Tragedy and mobilization:
  — Existential threats stimulate a social movement to transform multilateral cooperation, which disrupts economic incentives; enhances non-state actors' role.
  — Sustainable development is promoted along with China's new energy technologies, creating an unlikely partnership.
  — Global energy revolution creates a backlash by fossil fuel industry-based nations.
  — Nongovernmental organizations and multilateral organizations influence standards, resources, and so on and prod states to action.

The scenarios in the *Global Trends* report helped to inform the committee's scenario development, but, consistent with the statement of task, the committee identified drivers and examined scenarios focused on encryption.

E

Acronyms and Abbreviations

AES	Advanced Encryption Standard
AI	artificial intelligence
ANSI	American National Standards Institute
BSI	Bundesamt für Sicherheit in der Informationstechnik (German Federal Office for Information Security)
C-RAN	Cloud Radio Access Network
CAC	Common Access Card
CFIUS	Committee on Foreign Investment in the United States
CPU	central processing unit
CRQC	Cryptographically Relevant Quantum Computer
DES	Data Encryption Standard
DevOps	Software development (Dev) and IT operations (Ops)
DHS	Department of Homeland Security
ECDH	Elliptic Curve Diffie-Hellman
ECDSA	Elliptic Curve Digital Signature Algorithm
FHE	fully homomorphic encryption
FIPS	Federal Information Processing Standard
FSS	Function Secret Sharing
GCHQ	Government Communications Headquarters
GDPR	Union General Data Protection Regulation
GPU	graphics processing unit
GSM	Global System for Mobile Communications

APPENDIX E

HTTPS	Hypertext Transfer Protocol Secure	
IEC	International Electrotechnical Commission	
IETF	Internet Engineering Task Force	
IoT	Internet of Things	
IP	Internet protocol	
IPsec	Internet protocol security	
ISO	International Organization for Standardization	
IT	information technology	
KDC	key distribution center	
LTE	Long Term Evolution (usually called 4G LTE)	
MAC	method authentication code	
MPC	multi-party computation	
NFV	Network Functions Virtualization	
NISQ	noisy, intermediate-scale quantum	
NIST	National Institute of Standards and Technology	
NSA	National Security Agency	
ODNI	Office of the Director of National Intelligence	
OPM	Office of Personnel Management	
ORAM	Oblivious Random Access Memory	
PIR	Private Information Retrieval	
PKI	public-key infrastructure	
PQC	post-quantum cryptography	
PRNG	pseudo-random number generator	
PSI	private set intersection	
PSU	private set union	
QKD	quantum key distribution	
RFC	request for comments	
RSA	Rivest-Shamir-Adleman (the developers)	
SCADA	Supervisory Control and Data Acquisition	
SDN	Software-Defined Networks	
SGX	software guard extensions	
SHA	Secure Hash Algorithm	
SIKE	Supersingular Isogeny Key Encapsulation	
SSL	Secure Sockets Layer	
TLS	Transport Layer Security	

VCAT	Visiting Committee on Advanced Technology
VPN	virtual private network
WG	working group
ZK	zero knowledge

F

Committee Member Biographical Information

STEVEN B. LIPNER, *Chair*, is the executive director of SAFECode, a non-profit organization dedicated to increasing trust in information and communications technology products and services through the advancement of effective software assurance methods. Mr. Lipner is also an adjunct professor of computer science in the Institute for Software Research in the School of Computer Science at Carnegie Mellon University. Mr. Lipner retired in 2015 as the partner director of software security in Trustworthy Computing at Microsoft Corporation. His expertise is in software security, software vulnerabilities, Internet security, and organization change for security. He is the founder and long-time leader of the Security Development Lifecycle team that has delivered processes, tools and associated guidance, and oversight that have significantly improved the security of Microsoft's software. Mr. Lipner has more than 50 years of experience as a researcher, development manager, and general manager in information technology security. He served as the executive vice president and general manager for Network Security Products at Trusted Information Systems and has been responsible for the development of mathematical models of security and of a number of secure operating systems. Mr. Lipner was one of the initial 12 members of the U.S. Computer Systems Security and Privacy Advisory Board and served two terms and a total of 10 years on the board. He now serves as the chair of the board's successor, the Information Security and Privacy Advisory Board. Mr. Lipner is the author of numerous professional papers and has spoken on security topics at many professional conferences. He is named as the inventor on 12 U.S. patents in the fields of computer and network security and has served on numerous scientific boards and advisory committees, including committees of the National Academies of Sciences, Engineering, and Medicine, such as the Committee on Future Research Goals and Directions for Foundational Science in Cybersecurity and the Committee on Law Enforcement and Intelligence Access to Plaintext Information in an Era of Widespread Strong Encryption: Options and Tradeoffs. Mr. Lipner was elected in 2015 to the National Cybersecurity Hall of Fame and in 2017 to the National Academy of Engineering (NAE).

MARK LOWENTHAL, *Vice Chair*, teaches at the Intelligence & Security Academy, LLC, a national security education, training, and consulting company where he was formerly the president and chief executive officer. Dr. Lowenthal is an internationally recognized expert on intelligence. He is also on the faculty at the Krieger School of Arts and Sciences at Johns Hopkins University in Washington, DC. From 2002 to 2005, Dr. Lowenthal served as the assistant director of Central Intelligence for Analysis and Production and as the vice chairperson for evaluation on the National Intelligence Council. Prior to these duties, he served as the counselor to the director of Central Intelligence. Dr. Lowenthal has written extensively on intelligence and national security issues, including 6 books and more than 100 articles or studies. His most recent book, *Intelligence: From Secrets to Policy* (Sage/CQ Press, 8th ed., 2019), has become the standard college and graduate school textbook on the subject. Dr. Lowenthal received his B.A. from Brooklyn College and his Ph.D. in history from Harvard University and was awarded the National Intelligence Distinguished Service Medal, the Intelligence Community's highest award. Dr. Lowenthal is a former staff director of the House Permanent Select Committee on Intelligence.

HANS ROBERT DAVIES is a futurist at Toffler Associates. Mr. Davies specializes in strategic planning, risk management, resource management, and innovation policy. He helps organizations strengthen their ability to manage enterprise risk. Prior to joining Toffler Associates, Mr. Davies worked at SAIC, supporting arms control and complex operations initiatives for the Department of Defense. He has served as a congressional staff member, worked with the Department of State, and was a Robert Bosch Foundation Fellow in Germany. Mr. Davies earned a B.A. with honors in history from Williams College and an M.A. in international relations and international economics from Johns Hopkins University.

CHIP ELLIOTT holds more than 90 patents, primarily in computer networks, and is a fellow of the American Association for the Advancement of Science (AAAS), the Association for Computing Machinery (ACM), and the Institute of Electrical and Electronics Engineers (IEEE). Before retiring, Mr. Elliott served as a researcher and then as the chief technology officer for Raytheon BBN.

GLENN S. GERSTELL is the senior advisor at the Center for Strategic and International Studies. Mr. Gerstell served as the general counsel of the National Security Agency (NSA) and the Central Security Service from 2015 to 2020. He has written and spoken widely about the intersections of technology and national security and privacy. Prior to joining the NSA, Mr. Gerstell practiced law for almost 40 years at the international law firm of Milbank, LLP, where he focused on the global telecommunications industry and served as the managing partner of the firm's Washington, DC, Singapore, and Hong Kong offices. Mr. Gerstell served on the President's National Infrastructure Advisory Council, which reports to the president and the Secretary of Homeland Security on security threats to the nation's infrastructure, as well as on the District of Columbia Homeland Security Commission. A graduate of New York University and Columbia University School of Law, Mr. Gerstell is an elected member of the American Academy of Diplomacy and a member of the Council on Foreign Relations. Earlier in his career, he was an adjunct law professor at the Georgetown University Law Center and the New York Law School. He is a recipient of the National Intelligence Distinguished Service Medal, the Secretary of Defense Medal for Exceptional Civilian Service, and the NSA Distinguished Civilian Service Medal.

NADIA HENINGER is an associate professor in computer science and engineering at the University of California, San Diego. Dr. Heninger's research focuses on applied cryptography and security, particularly cryptanalysis of public-key cryptography in practice. She is the recipient of a 2017 National Science Foundation (NSF) CAREER award, and her research has won best paper awards at CCS 2016, CCS 2015, Usenix Security 2012, and a best student paper award at Usenix Security 2008. Previously, Dr. Heninger was an assistant professor at the University of Pennsylvania. She received her Ph.D. in computer science in 2011 from Princeton University and spent time as a postdoctorate at the University of California, San Diego, and Microsoft Research New England.

SENY KAMARA is an associate professor of computer science at Brown University and the chief scientist at Aroki Systems. Before joining Brown, Mr. Kamara was a researcher at Microsoft Research (Redmond Lab). His research is in cryptography and is driven by real-world problems from privacy, security, and surveillance. Mr. Kamara has worked extensively on the design and cryptanalysis of encrypted search algorithms, which are efficient algorithms to search on end-to-end encrypted data. He maintains interests in various aspects of theory and systems, including applied and theoretical cryptography, data structures and algorithms, databases, networking, game theory, and technology policy. Mr. Kamara also directs the Encrypted Systems Lab and is affiliated with the CAPS group, the Data Science Initiative, and the Center for Human Rights and Humanitarian Studies.

PAUL CARL KOCHER is an American cryptographer and a cryptography entrepreneur who founded Cryptography Research, Inc. (CRI) and served as its president and chief scientist. Mr. Kocher received a bachelor's degree in biology from Stanford University in 1995, where he worked part-time with Martin Hellman. As demand for Mr. Kocher's knowledge in cryptography escalated, he gave up on his original plan to become a veterinarian and founded CRI instead. Mr. Kocher pioneered the field of side-channel attacks, including the development of timing attacks that can break implementations of RSA, DSA, and fixed-exponent Diffie-Hellman that operate in non-constant time, as well as the co-development of power analysis and differential power analysis. His side-channel attack countermeasure designs are widely deployed in secure integrated circuits and other cryptographic devices. Mr. Kocher has also worked on microprocessor security, and co-discovered and named the Spectre vulnerability, which leverages speculative execution and other microprocessor performance optimizations to extract confidential information. He also helped architect security-related integrated circuits, including Deep Crack, a DES brute-force key search machine. Mr. Kocher is a member of the NAE.

BRIAN LaMACCHIA is a Distinguished Engineer at Microsoft Corporation and heads the Security and Cryptography team within Microsoft Research (MSR). His team's main project at present is the development of quantum-resistant, public-key cryptographic algorithms and protocols. Dr. LaMacchia is also a founding member of the Microsoft Cryptography Review Board and consults on security and cryptography architectures, protocols, and implementations across the company. Before moving into MSR in 2009, he was the architect for cryptography in Windows Security, development lead for .NET Framework Security, and program manager for core cryptography in Windows 2000. Prior to joining Microsoft, Dr. LaMacchia was a member of the Public Policy Research Group at AT&T Labs—Research. In addition to his responsibilities at Microsoft, he is an adjunct associate professor in the School of Informatics and Computing at Indiana University Bloomington, and an affiliate faculty member of the Department of Computer Science and Engineering at the University of Washington. Dr. LaMacchia also currently serves as the treasurer of the International Association for Cryptologic Research (IACR) and as a vice president of the board of directors of the Seattle Opera. He received an S.B., an S.M., and a Ph.D. in electrical engineering and computer science from the Massachusetts Institute of Technology (MIT) in 1990, 1991, and 1996, respectively.

BUTLER W. LAMPSON is a technical fellow at Microsoft Corporation and an adjunct professor at MIT. Dr. Lampson has worked on computer architecture, local area networks, raster printers, page description languages, operating systems, remote procedure call, programming languages and their semantics, programming in the large, fault-tolerant computing, transaction processing, computer security, WYSIWYG editors, and tablet computers. He was one of the designers of the SDS 940 time-sharing system, the Alto personal distributed computing system, the Xerox 9700 laser printer, two-phase commit protocols, the Autonet LAN, the SPKI system for network security, the Microsoft Tablet PC software, the Microsoft Palladium high-assurance stack, and several programming languages. Dr. Lampson received the ACM Software Systems Award in 1984 for his work on the Alto, the IEEE Computer Pioneer award in 1996 and von Neumann Medal in 2001, the Turing Award in 1992, and the NAE's Draper Prize in 2004. He is a member of the National Academy of Sciences and the NAE and a fellow of the ACM and the American Academy of Arts & Sciences.

RAFAIL OSTROVSKY is a Distinguished Professor of Computer Science and a Distinguished Professor of Mathematics at the University of California, Los Angeles (UCLA). Dr. Ostrovsky joined UCLA in 2003 as a full tenured professor, coming from Bell Communications Research, where he was a senior research scientist. Prior to beginning his career at Bellcore, he was an NSF Mathematical Sciences Postdoctoral Research Fellow at the University of California, Berkeley. Dr. Ostrovsky received his Ph.D. in computer science from MIT in 1992 (advisor: Silvio Micali; thesis: "Software Protection and Simulation on Oblivious RAM"), supported by an IBM graduate fellowship. Dr. Ostrovsky is a fellow of IEEE, a fellow of IACR, and a foreign member of Academia Europaea. He has 14 U.S. patents issued and more than 300 papers published in refereed journals and conferences. Dr. Ostrovsky served as a chair of the IEEE Technical Committee on Mathematical Foundations of Computing from 2015 to 2018 and has served on more than 40 international conference program committees, including serving as the chair of the Foundations of Computer Science 2011. He is a member of the editorial boards of the *Journal of the*

ACM, *Algorithmica*, and the *Journal of Cryptology* and is the recipient of multiple awards and honors, including the 2017 IEEE Computer Society Technical Achievement Award and the 2018 RSA Conference Excellence in the Field of Mathematics lifetime achievement Award. At UCLA, Dr. Ostrovsky heads the Center of Information and Computation Security, a multidisciplinary Research Center at the Henry Samueli School of Engineering and Applied Science.

ELIZABETH RINDSKOPF PARKER retired as the executive director of the State Bar of California. Ms. Parker previously served as the dean of the McGeorge School of Law at the University of the Pacific from 2002 to 2012. She was general counsel with the University of Wisconsin system from 1999 to 2002. Before that, Ms. Parker was general counsel of the Central Intelligence Agency from 1990 to 1995. She was also the principal deputy legal advisor to the Department of State from 1989 to 1990 and general counsel for the NSA from 1984 to 1989. Ms. Parker received her J.D. from the University of Michigan. She is a lifetime counselor and former chair of the American Bar Association Standing Committee on Law and National Security and holds membership in the American Bar Foundation and the Council on Foreign Relations. She has served on a number of committees at the National Academies. Ms. Parker is also a two-term presidential appointee to the Public Interest Declassification Board.

PETER SWIRE is the Elizabeth and Tommy Holder Chair of Law and Ethics at the Georgia Tech Scheller College of Business, where he teaches cybersecurity and privacy. Mr. Swire is senior counsel with Alston & Bird LLP, the research director for the Cross-Border Data Forum, and a member of the National Academies' Forum on Cyber Resilience. In 2019, the Future of Privacy Forum honored him for Outstanding Academic Scholarship. In 2018, Mr. Swire was named an Andrew Carnegie Fellow for his project on cross-border data flows. In 2015, the International Association of Privacy Professionals awarded him its Privacy Leadership Award. In 2013, he served as one of five members of President Obama's Review Group on Intelligence and Communications Technology. In 2009–2010, he served as the Special Assistant to President Obama for Economic Policy. Under President Clinton, Mr. Swire was the Chief Counselor for Privacy, the first person to have U.S. government-wide responsibility for privacy policy. In that role, his activities included being White House coordinator for the HIPAA medical privacy rule, chairing the White House working group on encryption, and helping negotiate the U.S.-European Union Safe Harbor agreement for trans-border data flows. Mr. Swire graduated summa cum laude from Princeton University in economics and public policy, and from the Yale Law School.

PETER J. WEINBERGER is a software engineer at Google, Inc. Dr. Weinberger is a computer scientist best known for his early work at Bell Labs. He was an undergraduate at Swarthmore College, graduating in 1964. He received his Ph.D. in mathematics in 1969 from the University of California, Berkeley, in number theory. After holding a position in the Department of Mathematics at the University of Michigan, Ann Arbor, where he continued his work in number theory, Dr. Weinberger moved to AT&T Bell Labs, where he contributed to the design of the AWK programming language (he is the "W" in AWK), and worked on database systems, and a Fortran compiler. Prior to joining Google, Dr. Weinberger was at Renaissance Technologies, a hedge fund. He is a fellow of the AAAS, and is on various committees giving technical advice to the U.S. government.